Plastics in Automotive Engineering

Plastics in Automotive Engineering

Materials – Components – Systems

Edited by
H.-G. Haldenwanger and L. Vollrath

Hanser Publishers, Munich Vienna New York

Hanser/Gardner Publications, Inc., Cincinnati

Distributed in the USA and in Canada by
Hanser/Gardner Publications, Inc.
6600 Clough Pike, Cincinnati, Ohio 45244-4090, USA
Fax: +1 (513) 527-8950

Distributed in all other countries by
Carl Hanser Verlag
Postfach 86 04 20, 81631 München, Germany
Fax +49 (89) 98 48 09

Library of Congress Cataloging-in-Publication Data

Plastics in automotive engineering : materials, components, systems /
edited by H.-G. Haldenwanger and L. Vollrath.
p. cm.
Includes index.
ISBN 1.56990-132-5 (U.S.)
1. Plastics in automobiles. I. Haldenwanger, H.-G. (Hans-Günther)
II. Vollrath, L.
TL. 154.P544 1994
629.23'2--dc20 94-4212

Die Deutsche Bibliothek – CIP-Einheitsaufnahme

Plastics in automotive engineering : materials, components,
systems / ed. by H-G. Haldenwanger and L. Vollrath. –
Munich ; Vienna ; New York : Hanser ; Cincinnati :
Hanser/Gardner, 1994
ISBN 3-446-17805-8 (Hanser)
ISBN 1-56990-132-5 (Hanser/Gardner)
NE: Haldenwanger, Hans-Günther [Hrsg.]

Printed in Germany

Preface

The automotive industry is facing fundamental changes in its structure. It is not only a question of the place of the car in society and its influence on the environment, economy and politics, but also of changes in technological fields: development, design, production and quality assurance. It is in these fields, in particular that the rules of the game for manufacturers and suppliers have been drawn into the slipstream of a general reorientation. A new trend has emerged, focussing on intensive cooperation with a smaller number of companies, which are no longer mere suppliers of individual parts are called upon to contribute where possible to the joint development of "systems".

These new closer ties between partners are part of the lean production system which grants a significant role to the suppliers in design, testing, production planning and logistics. As a result, the interaction between raw materials, components and systems and their environmental compatibility, necessary for such a product to be successful in the market, gains in importance as a result. This is where it becomes necessary to reduce the amount of raw materials and energy used in the manufacture, operation and final disposal of vehicles. Making an overall balance will facilitate the selection of materials and processes, while highlighting their future research and development potentials.

Furthermore, the overall balance with respect to the conservation of resources could be extended to include an ecological balance so that the environmental impact emanating from products, processes, technical systems or individual production units can be clearly identified, quantified and made the subject of a comparative evaluation. The identification of weak points or alternatives having better environmental compatibility could help in minimizing the impact of products and processes on the environment. A comparative evaluation, however, would reasonably presume that the consequences of the entire useful life, which are related to resources and environment, must be given due consideration.

Since product-oriented environmental protection will play an increasingly important role in future endeavours to protect the environment, it is particularly important for manufacturers and suppliers of entire systems, who must be aware of the increasing responsibility for their products beyond the limits of their engineering fields, to make a useful life balance as early as during the design stage of a product (definition stage). For it is the need to reduce vehicle weight, without loss of functionality and convenience, and to improve the efficiency of drive assemblies, that will clear the way for new lightweight construction concepts presenting modified materials and processes, i.e. the future is for intelligent combinations of materials allowing recyclable lightweight construction.

The content of this book is based on the most recent R & D findings of the European motorcar marcet; however, car manufacturing has grown into a global industry and its impact is the same everywhere on the worldwide community of car manufacturers, subcontractors, suppliers and the plastics industry too.

The editors are grateful to the authors for their comprehensive articles and for the willingness to provide hitherto unpublished results of their work. We are specially grateful to Monika Stüve for her neverending efforts to have this book published in time .

Hans-Günther Haldenwanger
Ludwig Vollrath

March 23, 1994

Contents

Life Cycle Analysis of car parts made of steel, aluminium and synthetics demonstrated on oil filters for car engines

Peter Eyerer, Jens Hesselbach, Oswald Part, Ingrid Pfleiderer, Konrad Saur, Manfred Schuckert

Introduction

The Life Cycle Analysis is an additional tool which contributes comprehensive information to to the decision-making process of engineering. If an engineer has the choice between several material alternatives and their specific processing, he can only make a successful decision if objective information is available to him. This includes in particular the specific histories of the materials to be used, the consideration of the alternatives in the utilisation phase and the requirements at the end of the serviceable life of the material.

Only the overlapping of technical, economical and ecological requirements will result in an "ideal" product, by taking a cycle with minimised resources into consideration. For that reason, a Life Cycle Analysis aiming for real results can only be carried out successfully on concrete parts or methods, [1] through [5].

Definition of a Life Cycle Analysis

The Life Cycle Analysis is a part-, method- and system-related instrument for the research of environmental parameters with the background of technical and economical specifications. For that purpose raw material, energy, emission, waste water and waste balances are carried out throughout the total life cycle. Boundary conditions must be precisely specified in each individual case. Figure 1 illustrates the dimension of this definition:
- specification: technical, economical, environmental,
- life cycle,
- balance flows: air, water, ground.

Product specification

In order to turn ideas into products, developers employ a multitude of instruments that have been tested and advanced for many years, Figure 2. Technical specifications are realised by means of various C-methods. Precalculation as well as follow-up calculation, value analysis, to name just two methods, show the economical degree of the realisation.
All instruments depicted so far in Figure 2 allow for active entrepreneurial interventions in the product development. From an environmental point of view of the engineer, the dominating factors are mostly requirements, regulations, laws, external obligations, i.e. reactive conditions.
In that connection, there is a demand for active, self-fashion-able aids. To this effect the Life Cycle Analysis assists in the creation, development, production, use and disposal of a product.

For that purpose it is strived to integrate the Life Cycle Analysis in the form of a Computer Aided Balancing (CAB) into C-methods that can be used for technical development. Under CAB we understand the computer-supported evaluation of environmental effects of given part alternatives based on a technical-economical specification. The section Software for Life Cycle Analysis examines this aspects in detail.

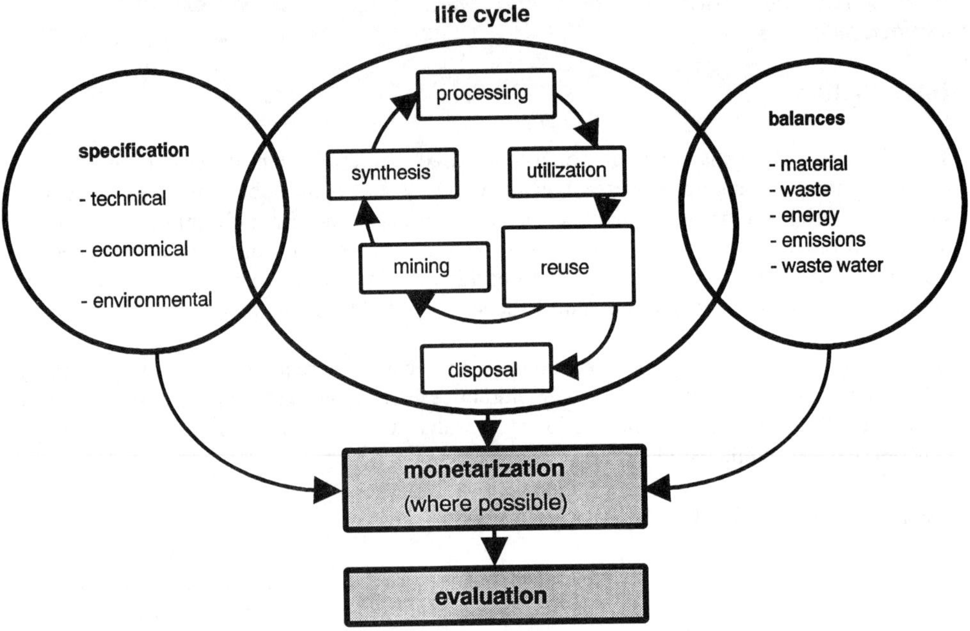

Figure 1: Life Cycle Analysis of parts, methods, products, systems

Life Cycle Analysis of parts - demonstrated on the oil filter

Initial situation

For almost 5 years the Institute for Polymer Testing and Polymer Science, University of Stuttgart, in collaboration with European industrial companies, has been developing methods of Life Cycle Analysis. In close cooperation with the PE Product Engineering GmbH, Ostfildern, a string of Life Cycle Analysis analysing different methods, parts and series or parts, could be successfully completed, Table 1. In that connection a three-year research project, in which over 40 German and European companies in the fields of steel, aluminium, polymers, manufacturers of machinery, suppliers, manufacturers of lacquers, waste disposal operations and car manufacturers participated, must be emphasised, Figure 3. The goal of the project was to establish the optimisation (strong and weak points) of Life Cycle Analysis on the example of car fenders of different designs (materials, methods).

The results can be found in a newly-developed software that was handed over to the project participants in a revised version at the end of November 1993.

A subsequent project of similar dimensions on the balancing of all kinds of recycling and disposal methods has been under way since September of 1993. Table 2 offers information on other current projects.

technical	a	b	c
project management			x
matreial data banks			x
construction method:			
CAE, CAD, CAM			x
work preparation (Refa)			x
quality assurance:			
QFD, CAQ, FMEA, SPC, BDE,			x
rapid testing, destructive and non-destructive tests			x
simulation methods (such as filled-in picture analysis)			x
quality audit such as evaluation of suppliers			x
production:			
CIE, CIM			x
simultaneous engineering			x
rapid prototyping (among others)			x

economical	a	b	c
internal:			
controlling			x
pre- and post calculation			x
external:			
value analysis (cost-value analysis)			x

environmental	a	b	c
environment official			x
operation guidelines			x
laws and regulations:			
* MAK, TRK			x
* bans: asbestos, Cd, FCKW			x
* waste, air, water junk cars, electronic scrap,		x	
wrappings			x
environmental compatibility test		x	
environmental audit	x		
evaluation of suppliers	x		
Life Cycle Analysis LCA	x		
CAB, CIB	x		

a in development b in installation c field-tested

Figure 2: Instruments for the evaluation and realization of specifications for technical products

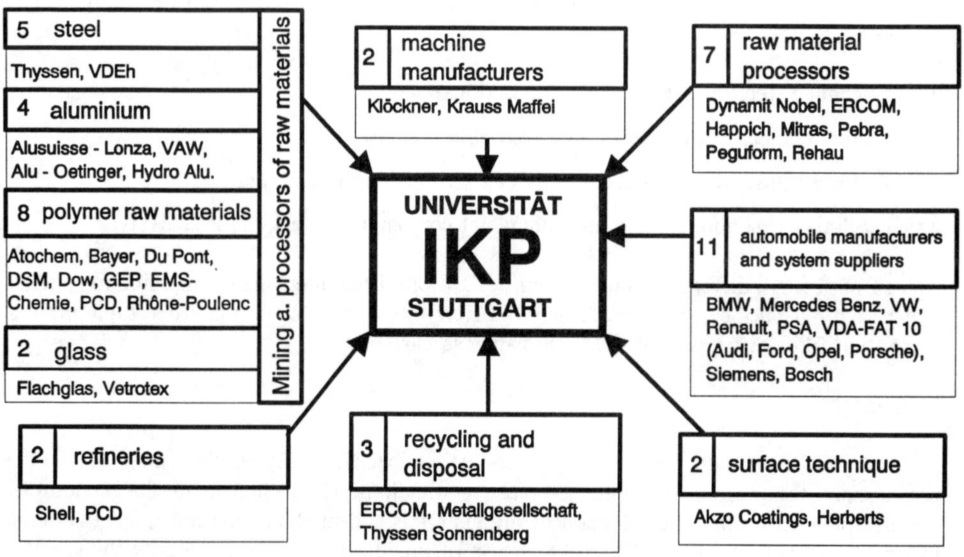

Figure 3: Life Cycle Analysis of automobile parts, "Fender" Project Group

Thanks to these intense projects with competent partners from the industries, in the meantime the most important factor of a balanced inventory, the data stock for many fields of application at the IKP, can be considered as being satisfactory, and even as very good in some areas, Figure 4.

Table 1: Completed projects on the Life Cycle Analysis

Completed projects	Materials / Processes
Car fender	Methodology- and software development; Steel, Aluminium, Polymere
Air intake- manifolds (3 constr.)	Secundary aluminium, Nylon 6.6 GF, multy-way-problem
Air filter	Steel, PP-T20
Oil filter	Secundary Aluminium, PA 6.6 GF, Multi- way problem
Use of varnish mud	Different processes
Car bumpers	Different Steels, Aluminium, PUR-S-RIM, GMT-PP, PP/EPDM, PC (PBT, PUR-RIM)
Commercial vehicle bumpers	Steel, SMC, PPO/PA

Life Cycle Analysis of three different oil filter designs

The primary consideration for the Life Cycle Analysis of different oil filter designs (car engines), i.e.

o easy-change steel oil filter (disposable casings)

o aluminium (diecast) oil filter casing (reusable system)

o Nylon 6.6 (glass fiber reinforced) oil filter casing (reusable system)

was a technical, economical and environmental comparison (weak point analysis).

Apart from certain differences due to their designs, all three filters are comparable in concept (filter cartridge surface), Figure 5. Figure 6 offers information about the processable masses (excluding filter cartridges) of the three balanced casings.

Method

In order to be able to carry out the production-specific analysis, the first step is the preparation of the technical and economical specifications. The technical and the economical specifications determine the selection of the materials and methods that are to be examined for the Life Cycle Analysis. This preselection of materials is then followed by the specific construction of the part.

After the different materials and the constructions resulting therefrom have been defined, the cycles from the supplying with raw materials to the reuse or recycling of the part or the material in question are determined. Alternative procedures steps within the cycles are taken into consideration.

After the material flows have been acquisitioned, mass and energy output of each individual method are determined (the mass flows also contain any emissions, recycling and waste flows). As a result, the analysis of the product cycle offers the cumulated data that is created during the use of the part (made of a variety of materials) throughout its serviceable life.

parts made of / balances	steel lit.	steel IKP	aluminum lit.	aluminum IKP	polymers packing	polymers technical polymers	polymers IKP
energy * without W + E * with W + E	+ -	+ +	+ -	+ +	+ o (-)	- - (x)	+ o
emissions	o	+	o	+	+	x	o
waste water	o	+	o	+	+	x	o
waste/residues	o	+	o(-)	+	+	x	+
landscape depletion	x	x	x	o	x	x	x
noise creation	x	x	x	x	x	x	x
data compiled in	91	93	91	93	93	80	93

+ updated data available
o data partially available
- data insufficiently available
x data missing

*W = reutilization
*E = disposal
Lit. ... data available to public
IKP ... data available to IKP

Figure 4: Published and IKP data on Life Cycle Analyses (status of 10/93)

Boundary conditions, system boundaries and data principles

The starting point of any Life Cycle Analysis must be a listing of the boundary conditions and system boundaries, Table 3.

Some relevant supplemental information:

serviceable life of an automobile: 150.000 km - 200.000 km
serviceable life of a cartridge: 15.000 km - 20.000 km

Easy-change steel oil filter:
10 oil filters (deep-drawn steel casings, Figure 5) as disposable products are required throughout the serviceable life of an automobile. To make a comparison with the technical

characteristics of the aluminium possible, respectively the synthetics filter, an aluminium filter top (reusable) must be taken into consideration in addition to the easy-change steel filter.

Aluminium oil filter:
Throughout the serviceable life of an automobile, one aluminium casing (reusable) and 10 cartridges (disposable) are needed.

PA 6.6 GF 35 oil filter:
Throughout the serviceable life of an automobile one PA 6.6 GF 35 casing (reusable) and 10 cartridges (disposable) are required. The synthetics version demonstrated in the following contains no metal screws, i.e. has been weight-optimized.

Currently system boundaries are still drawn often due to insufficient data availability, such as silicium as an alloying constituent for secondary aluminium, additives release agents for glass fibers, aggregates, etc.

Table 2: Current and planned projects on the Life Cycle Analysis of parts and systems

Projects in work	**Materials / Processes**
Structure components of cars: door, driver seat	Steel, Alu, Synthetics (SG, GMT, SMC, RRIM, SRIM, RTM)
Complete car	Systematic, important materials and processes
Mosaic (Renault)	Different materials
Car component: fuel- tank	PEHD (fluoride), Steel
Recycling processes	Recycling of particals, pyrolysis, alcoholysis, synthesis of Methanole without emissions, metallurgical total recycling
Air intake- manifold	SKT, 2-ST, and others
Planned projects	**Materials / Processes**
Building industry	Different building materials, raw materials, systems of buildings
electromech. components	Copper, other metals, Synthetics
Isolation of cables	Different Synthetics
Processes of varnishing	Water varnish, Powder varnish, LM- varnish

Even though some information exists on the material recycling of Nylon 6.6 GF, we balanced a 100 % disposal. That is in accordance with today´s situation in mass production. Now it already becomes apparent that a Life Cycle Analysis can never be more than a snapshot of concrete parts or methods, based on the existing supply conditions. General statements or deductions are not admissible in any case.

The process of a Life Cycle Analysis is generally the same for all parts, the individual data differs in each individual case. You start with the exploitation of raw materials, such as iron ore in Brazil for steel. In the case of the aluminium oil filter casing a secondary cycle of the aluminium had to be examined for the most part. 3 % loss of the mass was taken into considerations. 4 % of the mass is supplemented by primary aluminium. The collecting and processing of the aluminium scrap, in regards to the special situation of the producing filter manufacturer, was examined. Four suppliers with different data records, hauling distances, etc. had to be balanced.

Table 3: Boundary conditions for the Life Cycle Analysis of the examined oil filter casings

	Steel	Aluminum	PA 6.6 GF 35
Raw Materials	Exploitation analysed, search for raw materials not balanced		
Materialproduction	Route via converter-steel production no alloying elements considered	Mean of 4 paricular remelters taken, silicium considered energetically, outside of system: bondaries: primary aluminium production and additional alloying elements	outside of system: bondaries: lubricant cant for glass fibers, additives, catalysts (<3 per mil unit of mass) mining of colemanite, limestone, quartz only considered energetically
Processing	deep-drawing	diecasting	injection moulding
Treatment Small parts not considered	not required	considered (incl. washing etc.)	not required
Surface technique	varnishing examined	not required	not required
Utilization phase	x% liters reduced fuel consumption / y% weight savings and 100 km, only conclusion from altered CO2 emissions possible		
Disassembly, reutilization, **disposal**	Shredder Reuse in converter No cascade	Shredder Reuse in drum type furnace no cascade	100% disposal
Energy supply	recorded site-specific		
Analysis data	Industries 1991/1992; literature 1988-1992: filter values, energy demand, airborne and waterborne emissions (incomplete), waste, resource consumption		
Energy content of raw materials	___	Aluminum: 31.0 MJ	Crude oil: 42.0 MJ/kg natural gas: 46.0 MJ/kg
Coupled and by-products	main distribution factor: mass		
Types of waste	solid waste	Solid waste, sand, salt slag, high-aluminia residue	Solid waste, disposal of parts

The casting of the aluminium slugs was recorded at the sites of two suppliers to the filter manufacturing company. Data on their processing was collected with the aid of Mann & Hummel (Ludwigsburg). Information concerning the processing of PA 6.6 GF 35 into the oil filter casing also stems from Mann & Hummel. The glass fiber production was integrated into the examination with data from a glass fiber manufacturer, among other data.

Unfortunately the supplying polymer raw material processor was unwilling to cooperate. Any data concerning PA 6.6, however, were extracted from the latest technical literature, so that it must be assumed, that the actual requirements for the production of the synthetic do not differ substantially.

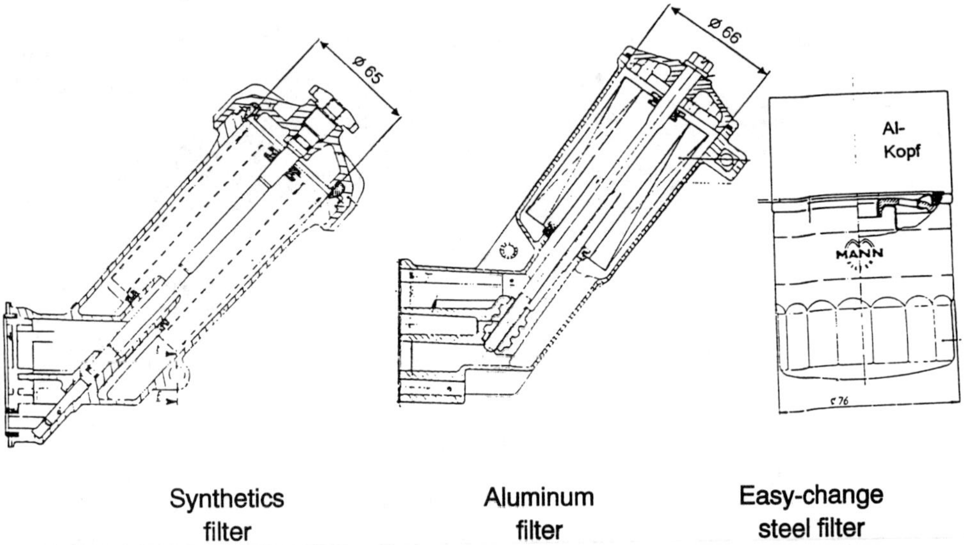

Figure 5: Geometrical sketches of examined oil filters

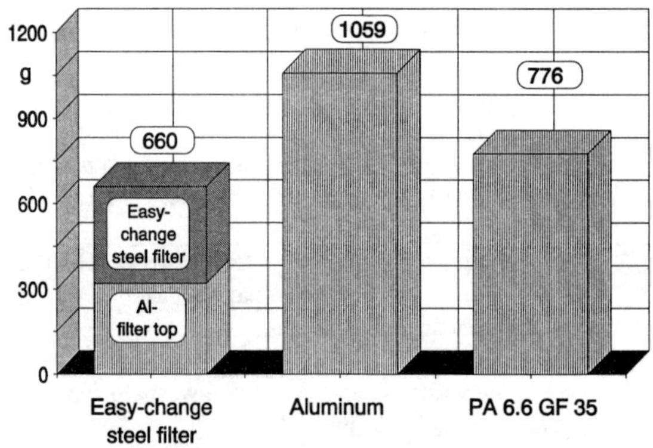

Figure 6: Processable masses (without filter cartridges) of compared oil filter casings

Material cycles

Steel filter casings with aluminium filter tops

Figure 7 depicts today's serial cycle of a disposable easy-change steel filter.

After the mining of iron ore, hard coal, etc., the smelting in the blast furnace and the subsequent production of crude steel, the required steel is produced in the mill. A galvanised coating reduces its corrodibility.

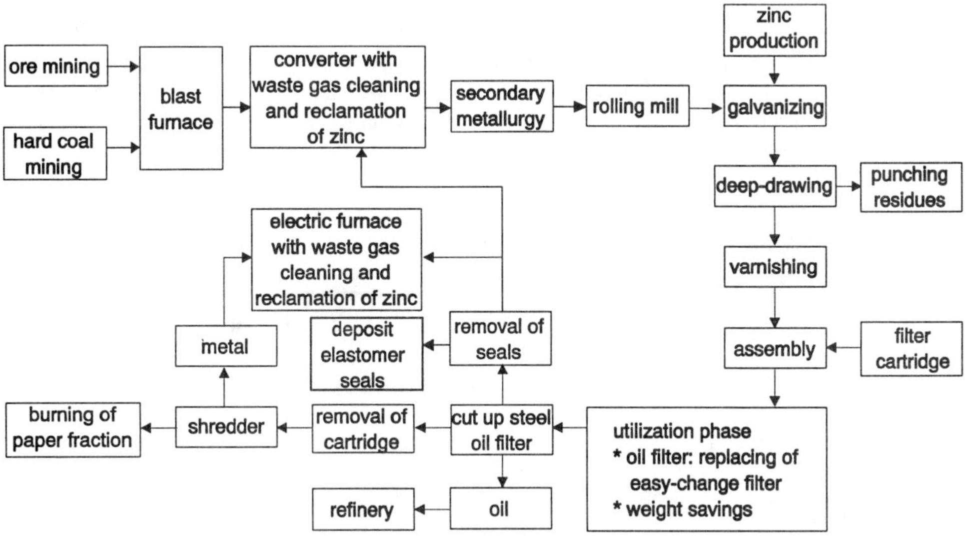

Figure 7: Life cycle of an easy-change steel oil filter

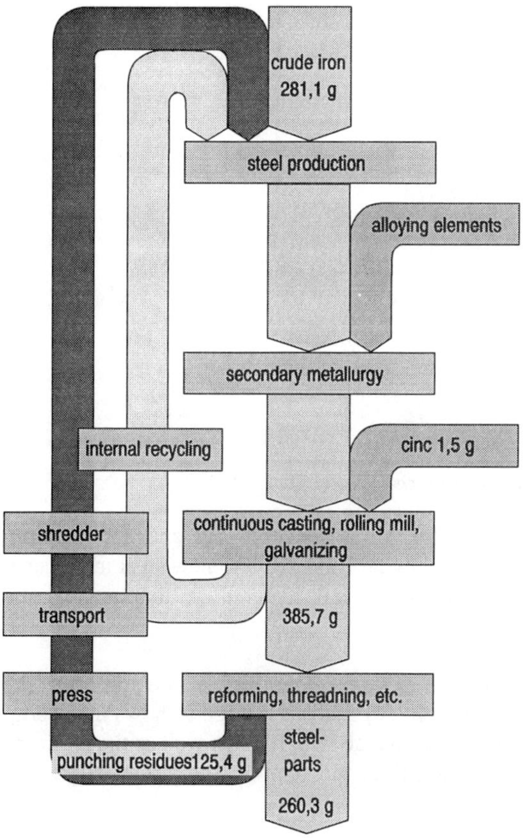

Figure 8: Production of a steel oil filter while taking recycling measures into consideration

The sheet steel pretreated in this manner is then deep-drawn. To protect the sheets further from corrosion, but also to satisfy aesthetic requirements, the manufactured pots are varnished. Then the filter cartridge is installed and the filter is sealed with seals as well as control systems. Within the framework of this analysis it is assumed that the filter (steel casing and filter cartridge) will be replaced ten times during the utilisation phase. The steel filter is cut open for the purpose of reuse and disposal.

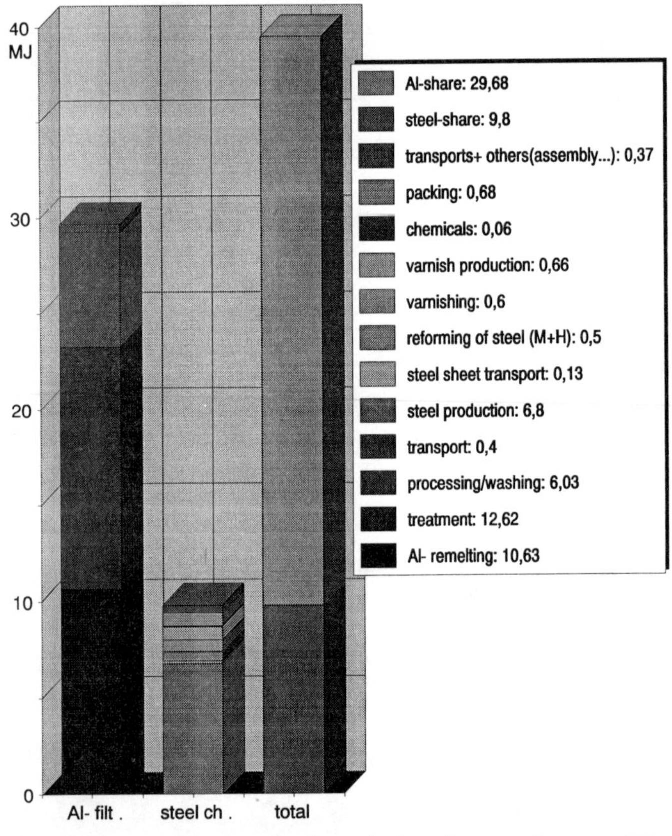

Figure 9: Energy requirements for the production of an easy-change steel filter

Afterwards the filter cartridges are taken out, the oil is squeezed out and subsequently disposed of in the incinerator for special waste. The metal is melted into new steel in the steel manufacturing process. The oil fraction of up to several hundred millimetres can be reprocessed in the refinery.

With a partial section of the material cycle for steel as an example, Figure 8 demonstrates the procedure for energy (Figure 9) and emission (Figure 10) analysis. Figure 9 shows the energy consumption required for the production of the easy-change steel filter as well as the pertinent aluminium filter top. It becomes apparent that the filter top requires the largest share, i.e.

approximately 75 % of the overall energy consumption. The easy-change steel filter per se consumes only about 10 MJ of primary energy equivalent.

The emissions shown in Figure 10 proportionally stem from the steel production, the aluminium diecasting for the filter top, the filter paper production (paper carbolized), the power supply, the processing of the parts and the varnishing.

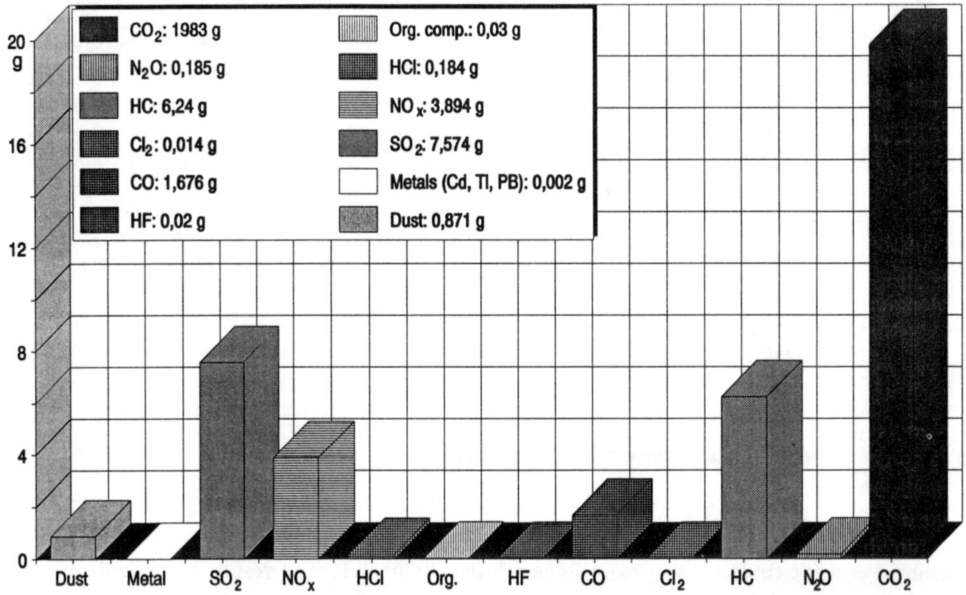

Figure 10: Atmospheric emissions during the production of the basic materials and their processing into an easy-change steel oil filter (incl. aluminium filter top)

Aluminium diecast oil filter casing

As initially mentioned, aluminium oil filter casings are mainly produced from secondary aluminium. Primary aluminium is mentally added with 4 %. Figure 11 shows the material cycle for primary and secondary aluminium.

The analysis of secondary aluminium begins with the collecting of aluminium scrap, takes their remelting in the drum type furnace, Figure 11, into account, then investigates the manufacturing of the casing in the diecasting method as well as the additional parts. Subsequently the utilisation phase as well as the recycling (shredder, collecting) are included in the investigation.

Including the remelting of aluminium scraps (the scraps are considered "free of energy" here, since the oil filter casing can technically be turned into a casing again after the utilisation phase; however; the loss of energy through the melting loss of aluminium is contained in the energy consumption with 3 %), the transport of the aluminium to the foundry and, following that, to the filter factory, the processing of the raw casing made of secondary aluminium uses 65.3 MJ of energy.

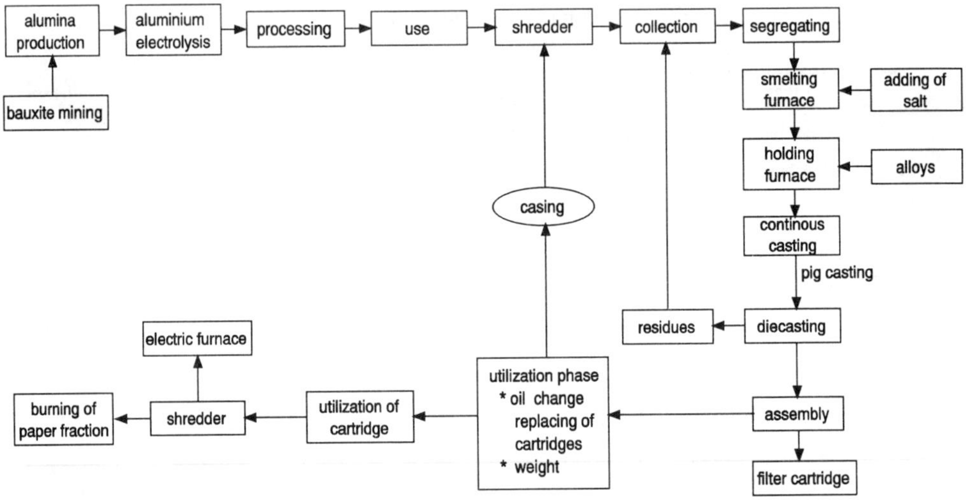

Figure 11: Material cycle of aluminium diecast filter casings

Figure 12 demonstrates the water pollution during the production of an aluminium oil filter casing. Pertinent findings also exist for steel and Nylon 6.6. The results are found in the conclusive evaluation.

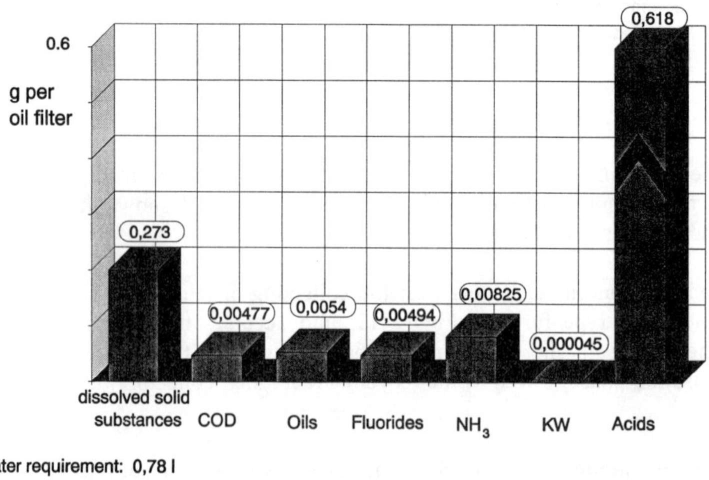

Figure 12: Water pollution during the production of an aluminium oil filter

Filter cartridge (paper)

The balance of the impact of the paper production on the environment begins with the logging and continues through the processing methods of fibrous material into paper. Figure 13 illustrates the production process of different paper and cardboard qualities. The filter paper is subsequently impregnated with phenolic resin and reinforced.

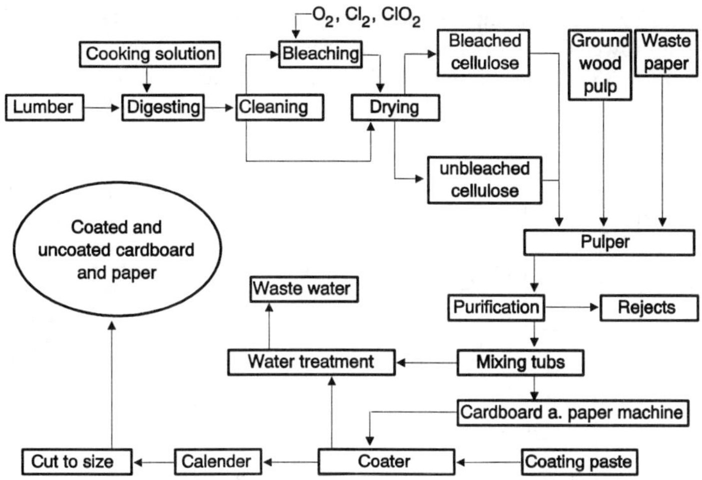

Figure 13: Procedure for the production of different paper and cardboard qualities

Table 4 shows the individual energy consumption for the processing of the basic materials and the production of a filter cartridge. There is a total energy requirement of 3.44 MJ per filter cartridge. Due to their extent, emissions and water pollution cannot be depicted; they are, however, included in the summary. 10 filter cartridges are needed per serviceable life (utilisation phase) of an automobile.

Table 4: Energy requirement for the production of the basic materials and their processing into a filter cartridge

Steel production	1,72 MJ
Paper production	0,92 MJ
Phenolic resin	0,20 MJ
Cartridge production	0,50 MJ
Reformed parts	0,10 MJ
Total energy requirement for a filter cartdidge	**3,44 MJ**

Oil filter casing made of Nylon 6.6 GF 35

For the production of a car oil filter made of PA 6.6 GF 35, Figure 14, the synthesis of PA 6.6 from hexamethylene diamine and adipic acid is at the focal point of all considerations. Subsequently the production of the raw materials, such as colemanite, limestone, caoline and silicium, into glass fibers is elucidated. After the compounding follows the injection moulding into the oil filter casing. During the utilisation phase the main emphasis is placed on the examination of the weight differences as compared to the other investigated oil filter versions and the savings in fuel with lower CO_2 emissions. At the end of the serviceable life of an automobile, a used PA 6.6 GF 35 oil filter casing would currently either be disposed of on the dump site or in the incinerator. For the future the issue of recycling of the synthetic must be solved, either in another oil filter casing or in a different use.

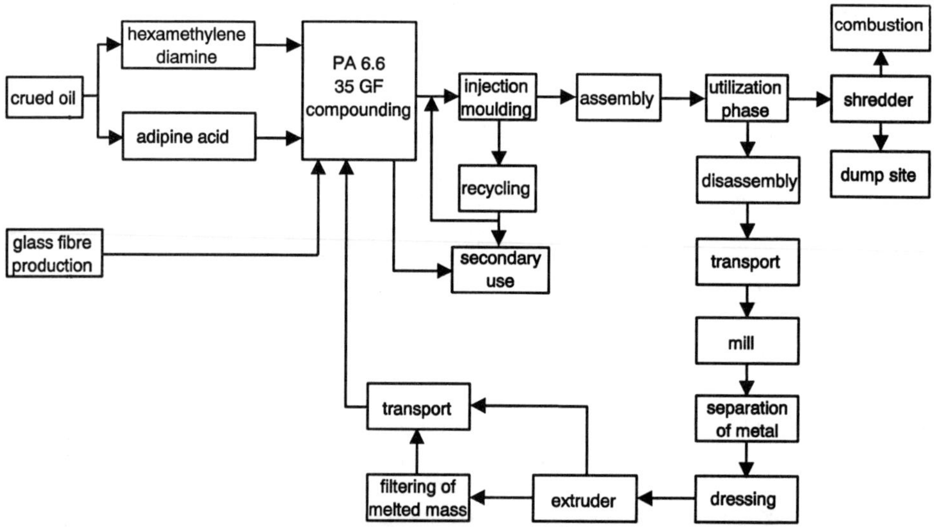

Figure 14: Potential life cycle of PA 6.6 GF oil filter

When examing the energy requirement for the production of the PA 6.6 GF 35 oil filter, two partial areas must be pointed out: The material production and the processing of the material into the oil filter. The production of an oil filter made of PA 6.6 GF 35 requires a total energy equivalent of 99.8 MJ. This includes the energy requirement for the synthesis of the PA 6.6 from crude oil, the production of glass fibers, the transport of all raw materials, the processing of PA 6.6 GF 35 into the finished oil filter at the filter factory, as well as the transport of the oil filter to the car manufacturer. Not included is the production of any catalysts during the production of PA 6.6, the additional materials for compound and small amounts during the PA 6.6 synthesis that lie within the per mil scope. 80 % of the energy requirement was used for the synthesis of PA 6.6 and the glass fiber production, 15 % for the injection moulding process, the remaining 5 % must be made available for the transport of the materials and of the part.

In this connection it should be noted that the value of 5 to 15 MJ for the injection moulding, as published by Kindler/Nikles [6] is not applicable to this case at all, since the process in question is considerably more complicated. The value for PA 6.6 GF 30 published at the time also could not be used for this study, because the IKP found different values during its research and detailed calculation.

Production and use of different oil filter designs

In the following an analysis will be carried out for three oil filter designs throughout a complete serviceable life. It is again emphasised that the specific conditions of Mann & Hummel Company, including their material suppliers, were investigated. Especially concerning the material PA 6.6 GF 35, there would be different results if other material suppliers had been selected (see below).

Approximately 43 MJ of primary energy equivalent are required for the production of an easy-change steel filter. This is somewhat greater than that of the comparable cartridges of the aluminium and synthetic oil filters. Less energy is used for its production due to its more advantageous geometry and manufacturing methods (3.44 MJ as compared to 4.13 MJ for the two other designs).

The share steel has in the energy consumption for the easy-change steel oil filter amounts to 9.8 MJ. This value includes the total material production, the transport energy, the forming energy, the varnishing, the computation of the requirement for the production of the varnish as well as the assembly and the packing. Therefore the essential parameters for this production are included, Figure 9.

Taking the ten replacements of the easy-change steel oil filter, including its cartridge, into consideration, the energy requirement for the steel design throughout the serviceable life of the automobile is increased from about 43 MJ to 162 MJ.

Comparing the aluminium oil filter, about 65 MJ of energy are used for the production of its casing and the additional parts. If you add all 10 cartridges to the production of the aluminium filter, you get approx. 107 MJ. The synthetic version with 10 cartridges requires approx. 141 MJ of primary energy. These values always include the complete material production, the casting, respectively injection moulding process, and the requirements for transport, assembly and packing.

It must be emphasised, however, that the route of synthesis of the synthetics is of great importance for the synthetics version, so that the values for the synthetics oil filter only apply to the route via hexamethylene diamine made of adipine acid.

When furthermore incorporating the energy consumption during the utilisation phase, represented by the reduction in fuel consumption caused by a reduction in weight, Figure 15 shows distinct differences in energy for an automobile of the upper level of medium-sized types of vehicles (approx. 1.300 kg of mass). The easy-change steel oil filter shows the lowest energy consumption (about 308 MJ), followed by the PA 6.6 version with about 312 MJ. Because of its heavy weight and the long utilisation phase of the oil filter the aluminium oil filter falls short by far (about 340 MJ). After all, the difference between different designs is 32 MJ. That is the approximate equivalent of the energy content of three-quarters kg of crude oil.

However, when taking 30 million cars in the Federal Republic of Germany or about 400 million cars world-wide into consideration, the optimisation of a part offers considerable saving potentials.

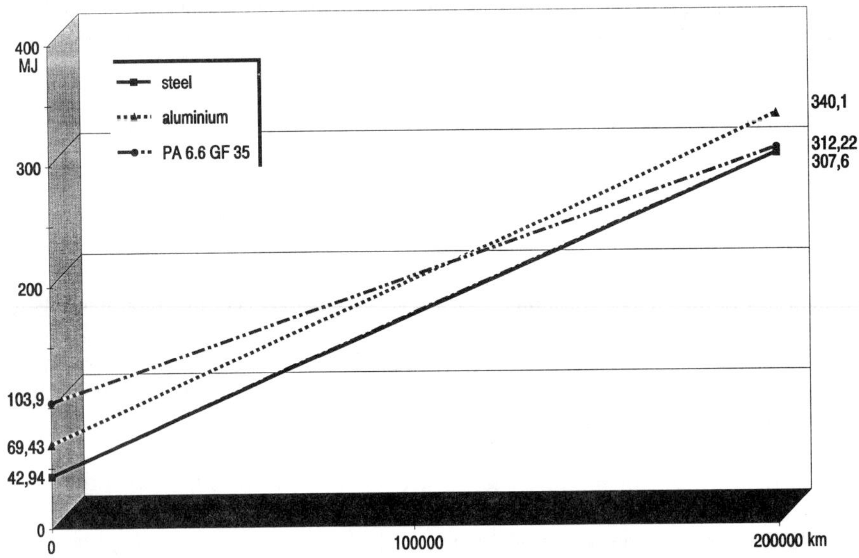

Figure 15: Energy consumption during the production and use of different oil filter versions in a BMW 320i (1315 kg, third mix 9,5 l/100 km, 200.000 km, 4,5 % rule)

Atmospheric emissions resulting from the production of oil filter casings

Figure 10 on the steel oil filter illustrates atmospheric emissions resulting from production. The emissions created when the energy is supplied are included and are usually the main cause of the pollution. For the processing of the supply situation for polyamide 6.6 GF 35 three production routes of three different raw material suppliers had to be balanced. The most important initial materials for the polymerization of PA 6.6 are adipine acid (AS) and hexamethylene diamine (HMDA). All three suppliers produce adipine acid the same way:

steam-cracker → benzene → cyclohexane → adipine acid

On the other hand, there are three routes for HMDA, all of which result in adiponitrile:

Route 1: steam-cracker → butadiene }
natural gas → HCN } → adiponitrile

Route 2: Crude oil (propylene) + natural gas (NH_3) →
acyrlonitrile → adiponitrile

Route 3: adipine acid + NH_3 → adiponitrile

This shows that Route 3 uses the most adipine acid, first for adipine acid itself, as Routes 1 and 2 do too, and then for adiponitrile. When nitrous oxide (N_2O) is emitted during the production of adipine acid, most will be emitted on Route 3. It must furthermore be assumed that Producer 3 will emit most of the created N_2O into the atmosphere, while Producer 2 will oxygenate the already smaller amount of N_2O.

So Producer 3 creates approx. 57.6 g of nitrous oxide for the PA 6.6 oil filter version. A GWP (Greenhouse Warming Potential) of 240 is assumed for nitrous oxide which, if converted into CO_2, means 13.82 kg of additional pollution for polyamide 6.6 of Producer 3 as compared to the steel or aluminium versions. This example effectively shows of how little help an Europe-wide or U.S.-wide or world-wide data average of the most diversified raw material producers is to the user. Weak point analyses would be hidden by it. The main advantage of a Life Cycle Analysis would be lost. An economical useful tool would consciously and intentionally be kept dull. The other atmospheric emissions during the production of polyamide 6.6 have been determined and are included in the evaluation. When comparing the values, one recognises that the easy-change steel filter shows the least emissions in regards to SO_2, NO_2, CO, HC and CO_2. After that the order, however, becomes uneven. While the aluminium oil filter shows less emissions in the area of NO_2, N_2O, HC and dust than the synthetics oil filter, the latter shows better results concerning CO, HCl, Cl_2 and HF.

CO_2 emissions during the production and utilisation of oil filter casings

A calculation of the CO_2 emission during the combustion of gasoline according to Birnbaum [7] shows that approx. 3.15 kg of CO_2 are emitted per kg of gasoline. When incorporating the operating ratio of the gasoline production as well as the CO_2 emission during the production and utilisation of different oil filter designs in an automobile.

The order does not change in relation to the energy consumptions as long as the greenhouse effect of N_2O is not taken into account. However, if the greenhouse effect of N_2O which is 240 times that of CO_2, is taken into consideration, this will result in drastic changes. Therefore the synthetics version has the greatest effects on the greenhouse. Certainly this statement only applies if a synthesis route is taken that actually shows the N_2O emissions, resp. if no N_2O retaining measures are taken.

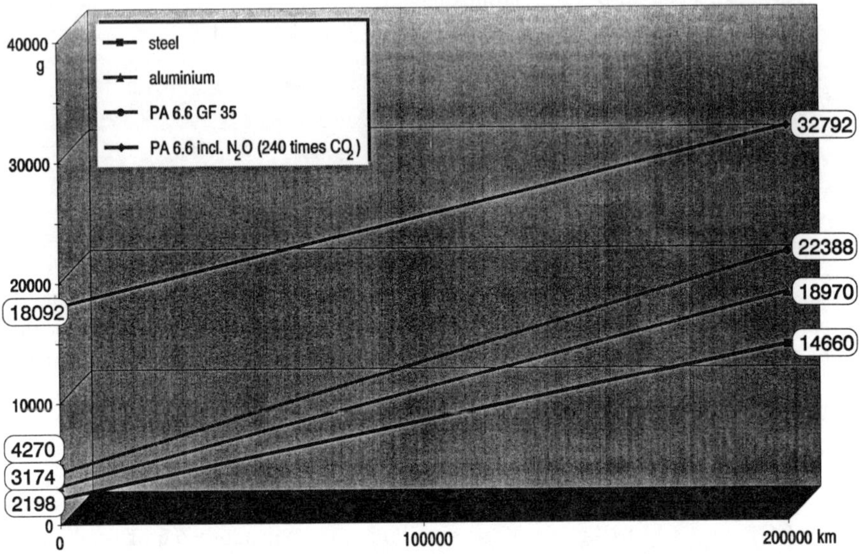

Figure 16: CO_2 emissions during the production and use of different oil filter versions in a BMW 320i (1315 kg, third mix 9,5 l/100 km, 200.000 km, 4,5 % rule)

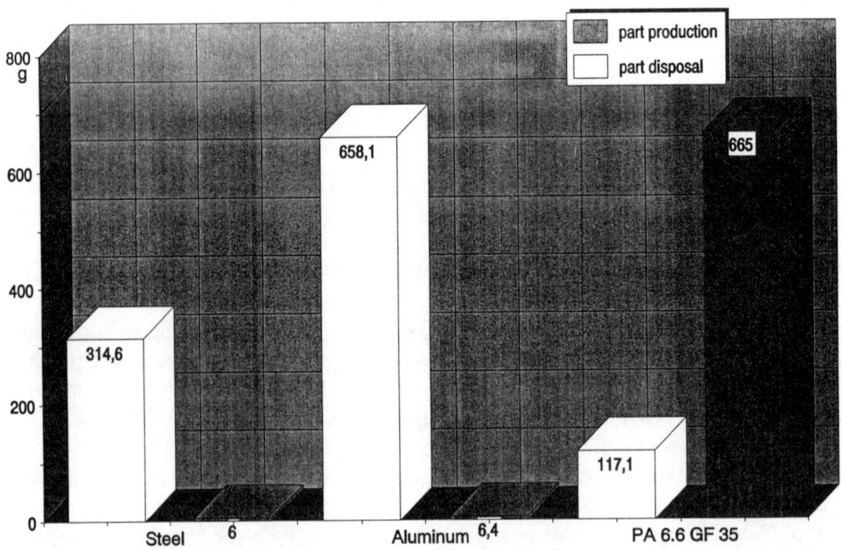

Figure 17: Dumping space requirements for the production and disposal of the different oil filter versions (without considering cartridges)

Disposal requirements for the production and disposal of oil filter casings

While the steel and aluminium version create the greatest disposal requirements through the production of parts and, when disposing of the parts, make rather good use of existing recycling measures and therefore use waste disposal space sparingly, the synthetics versions expose their weakest point especially when it comes to the latter - at least from the current point of view, Figure 17. Should a dismantling of larger synthetic part series occur in the future, the removal of synthetic oil filters - if in effect they had been used - would be very likely.

Due to the partial disposal of salt slag the aluminium oil filter requires the greatest requirement of disposal space for the part production. For the future a lesser disposal requirement for salt slag is to be expected, particularly in Germany, because recycling capacities in this field are currently being created. A further investigation of the disposal requirements created by the disposal of the cartridges reveals that, in comparison with the synthetics and aluminium versions, the cartridge for the easy-change steel filter is at an advantage due to its lesser weight. The former, on the other hand have a small advantage only in regards to the packing.

While it is currently unknown which quality characteristics the PA 6.6 GF 35 material, thus gained, would still have, a recycled admixture of 10 - 20 % to other, less polluted parts, would be conceivable.

Software [8], [9]

Requirements and limits of a software system on Life Cycle Analysis

While the strengths of data processing lie in the data storage and the process of formal algorithms (described in the so-called optimisation), programs are able to a rudimentary extent only to show a behaviour that approaches human intention or interpretation. For that reason, the user or the Life Cycle Analysis should take part in the selection of criteria as well as in the deductions from the results, as long as the criteria - what and in what way should it be analysed - have not yet been standardised. For the purpose of the program, however, attention must be paid to detailed information concerning the boundary conditions on which everything is based. It is this splitting of the preparation and the interface resulting therefrom that creates the necessity of user manual. In compliance with the terminology for product development and production this type of balancing can be called computer aided balancing (CAB).

The goal of CAB is a comparison between the ecological effects of different product alternatives that all fulfil the same given technical specification. Based on a rough product specification, a specification that is technically and ecologically intended can therefore be prepared, while taking the results of CAB into consideration.

Therefore CAB represents a useful component within the CIM that delivers additional criteria helpful for the selection during the phase of product planning in the light of the environmental applicability of the products to be selected from. A second essential field of application of CAB is the preparation of sensibility analysis formulas and scenarios.

Concept of the software system developed at the IKP, University of Stuttgart

Due to their great diffusion and reasonable price the range of IBM PC compatible hardware and software was suitable for such a system. In order to meet all requirements for the convenience of the user, the system is based on the MS Windows operating system extension and is equipped with the standard characteristics (menu, operating with the mouse, dialogue masks, context-related on-screen help, etc.).

Next to the desire for the greatest transparency possible during the analysis as well as during the subsequent, possibly later interpretation of data already collected was the demand for the user's opportunity to continuously expand the data stock. In order to fulfil both of these tasks and furthermore to be as soon as possible for future requirements, a modular approach geared towards the subject was made [8], [9].

The data subjects were divided into two directions:

- o horizontally into different parametric sections
- o vertically into hierarchic levels, which means that objects of greater complexity can be formed out of the chaining of simpler objects on the level below and, in turn, can subsequently be used an initial objects for additional, even more complex objects.

Structure of the user surface

Starting on the central survey screen, the user can branch out into all parts of the program. Vice versa he can return to said screen at any time during the program flow. At the centre of this mask is the icon for the calculation of the output lists, respectively values. Pictographs are arranged in the outer sections that correspond to horizontal hierarchic groups, respectively that belong into the section of "tool" which makes a series of useful operations available to the user that cannot be exclusively incorporated in one of the other sections.

The application of the illustrated concept shall now be explained in the light of the process technology field. Here the initial objects are flows (collective for substance and energy flows) as well as processes. These objects are created, resp. altered, by masks. A certain class, which can be referred to in another place, such as during the preparation of a process chain, is specified by this input. The subject of the next higher hierarchic level is the plan that describes a process chain in the form of a flow chart.

Plans are created with the help of a graphic editor. For that purpose universal process components are created at first that are only installed when reference to an existing process class has been made. Subsequently these processes can be linked with instances of flow classes also determined in advance into process chains.

The group of plans consisting of all plans that describe the life cycle of a part (plans for production, use and further treatment) is located on the highest hierarchic level of this partial section. Here again the plans described above represent classes that are installed into one or several groups of plans.

Summarised review of the software GaBi-Basic

Required hardware:

- o PC-AT 386/486
- o a minimum of 8 MB main memories
- o a minimum of 20 MB available hard disk space
- o graphic card and colour monitor for 600 x 800 dot definition
- o mouse

Required software:

- o MS-DOS 5.0
- o MS-Windows 3.1

Data contents:

- o Approximately 350 processes of the chemical, metal and synthetics producing and processing industries, including transport, with approx. 500 different substances and energy flows; a variety of process chain for the description of whole procedures.
- o The user can expand the data basis for as many processes as he chooses, and can create interactively additional process chains with the help of a graphic editor.

Evaluation of the Life Cycle Analysis as demonstrated on an oil filter casing

Object balances require evaluating summaries in order to reach decisions. Similar to object balances, evaluations are also based on temporal, local, socio-political and other boundary conditions. Given evaluation standards and methods therefore involve the danger to more or less interfere with the necessary freedom to make management decisions.

In line with the Life Cycle Analysis (object balance) [1] to [5] the method of a Life Cycle Analysis is to be developed which is an aid for the decision-making process but not a "handcuff" limiting the scope of managerial actions.

A socially accepted evaluation of ecological aspects of industrial products, methods or services that is subject to the most versatile claims made on such an evaluation is not realizable at this point in time. The claims made on the evaluation are too conflicting. Furthermore a justification based on scientific evidence on individual environmental aspects of specific emissions, waste or certain valuable substances is currently not available. (As an example it is herewith referred to the discussion on what greenhouse potential carbon dioxide CO_2 itself has. It is especially remarkable that the greenhouse potential of all substances is referred to this yet unknown greenhouse potential.)

And yet strategic operative decisions on management level in particular require evaluations. As long as an evaluation of monetary or non-monetary kind generally accepted by society is not available, evaluation of services, methods or products will have to be realised according or one's own standard. Suitable methods for that purpose must be developed.

Short description and discussion of existing evaluation methods (selection)

The result of the inventory which ideally will cover any aspects of the investigated product that are relevant to the environment, represents the data collected throughout the whole life cycle of the product. The goal of Life Cycle Analysis is the evaluation of, respectively the comparison between, different products or methods (functional equivalent) in reference to their environmental compatibility. This is usually only possible with the help of evaluation models, in which the basis data can be standardised. Thereby the selection of the evaluation model can influence the result considerably. Evaluation approaches and corresponding evaluation values are described in the following.

BUWAL Model of the "Ecological Shortage" [10]

For each harmful substance a certain equivalent value, the ecological factor, is determined. This ecological factor is calculated from the relation between the actual Impact F determined in the inventory and a maximum admissible Impact Fk, which barely stops short of causing irreversible damages to the ecological system. By multiplying this relation with the scale factor c, which has been determined for any emission and which indicates the sensibility of the ecological system towards said emission, you receive - after the standardisation with the critical impact - the ecological factor of a certain emission. The emission with the highest number of so-called ecological points strains the environment the most:

$$\textit{ecological factor} = \frac{1}{Fk} \cdot \frac{F}{Fk} \cdot C$$

A problem of this method is the determination of the critical impact and the scale factors. It is assumed that there is always an admissible amount of harmful substances that may be emitted. These values are political and therefore subjectively influenced, and currently there is no really clear scientific evidence available to justify them. Therefore this method has only an index, the non-dimensional ecological point, as its evaluated value. The critical flow, on whose determination the evaluation depends, serves as the evaluation standard.

Model ETH Zurich "Critical Impact" [11]

With this method emissions are weighted, while imission boundary values are taken into account, by dividing the emissions determined in the inventory by a specific boundary value. Hereby the volume of the emissions emitted into the water, air, resp. ground, which corresponds to the maximum admissible concentration per quantity of medium, is computed. The "critical impact" emerges after the division of the quantity of harmful substances determined in the inventory in a boundary value, such as the MIK or MAK value. These partial volumes are then added for the groups of ground, air and water.

Due to the great variation range and uncertainty of the boundary values the differences within the balance data are often exceeded, which can result in the fact that the selection of the boundary values can be decisive for the results of the analysis. The evaluation is done in the unit quantity of harmful substance per polluted air (reps. water or ground) per critical quantity. Concretely, relative to individual substances, this procedure really is suitable. As a cumulative value for diverse substances, however, it is questionable, since, among other things, reciprocal effects between different emissions cannot be taken into consideration. In addition to that the evaluation of emission based on dimensionless imission values must be viewed with concern.

VNCI Approach, Leiden, Netherlands [12]

Again based on the emissions determined, references are made to 12 different environmental impacts. These are as follows:Greenhouse potential, ozone damaging potential, acidification, eutrophication of waters, spreading of toxic substances, waste disposal, damaging of landscape, consumption of biological resources, deterioration of soil quality, loss of manifoldness of the biological, resp. cultural life.

For all emissions determined those shares concerning these 12 categories are determined that fall into those categories within a year, and are brought into relation with the overall emissions. This results in a percentile share of specific impacts of an analysed product on the environmental effects of a whole national economy-apart from the difficulty of classifying the often subjective criteria in amore exact way, a determination of the effects backed by natural science findings is problematic. Furthermore emissions also occur transnationally, so that, due to the entanglement of a single national economies of today, a reasonable attribute to individual products is not possible.

EPS System, Swedish Environmental Research Institute [13]

Here the environmental effects of 1 kg of a certain substance are the basic reference. This effect is called ELU (environmental load unit). The value of the environmental effects results from multiplication of value with the actual emission quantity. This value has been named ELV value (environmental load value). The ELU values are also determined for any manufacturing, service and transport processes. By taking a weighted avoidance cost approach, they result from multiplying the factors of relative avoidance costs per kg, share of the specific emission in the overall quantity emitted by the national economy. dwelling time in the area, the frequency of the emission, the spreading of the emission and a general characteristic value for the classification of the emitted substance itself. A problem in this connection is the determination of these individual factors which are mostly subjective or can only be determined with great difficulty.

Modified Efficiency-Cost-Analysis combined with 3D Portfolio method for the Life Cycle Analysis

The Life Cycle Efficiency-Cost-Analysis (GNKA), combined with the 3D Portfolio, is practice-oriented, systematic, easily modifiable in quality and therefore individually applicable.

It builds on the three technical, economical and environmental dimensions and can be illustratively depicted. The managerial responsibility and freedom to make decisions are fully retained, since all boundary conditions and emphases are determined by the company itself. The GNKA and the 3D Portfolio Method are versions of successful analyses that for decades have been applied to efficiency calculations and strategy issues in enterprises, [14].

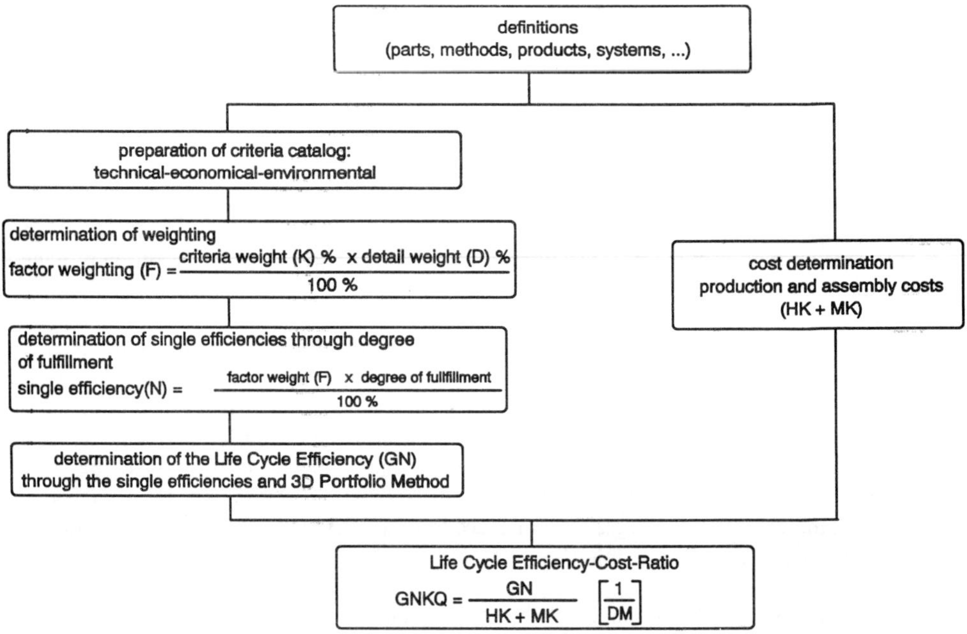

Figure 18: Flow chart of a Life Cycle Efficiency-Cost-Analysis (GNKA) with 3D Portfolio presentation

Criteria, weightings and evaluations

The GNKA and 3D Portfolio Method for the Life Cycle Analysis are demonstrated on the three oil filter designs. Figure 18 shows the procedure of a Life Cycle Efficiency-Cost-Analysis (GNKA). Subsequent to the definition of the versions to be examined (parts, methods) a catalogue of criteria according to technical, economical and environmental aspects is prepared. Figure 19 summarises the required partial steps and indicates the corresponding weightings.

These are:

$$factor\ weight\ (F)\ \% = \frac{criteria\ weight\ (K)\ \% \cdot detail\ weight\ (D)\ \%}{100\ \%}$$

and the

$$individual\ effenciency\ (EN) = \frac{factor\ weight\ (F)x\% \cdot performance\ degree\ (e)\ \%}{100\ \%}$$

The performance degree E reaches from 1=poor to 5=very good.

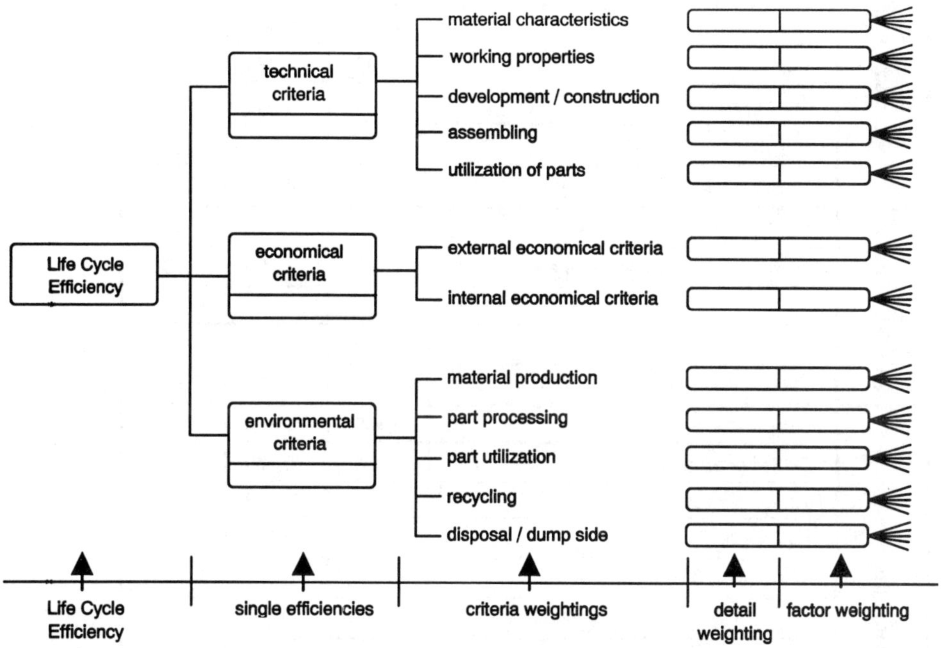

Figure 19: Partial steps to determine a Life Cycle Efficiency from single efficiencies and the required weightings

Table 5 summarises all determined individual efficiencies and allows for single comparing evaluations or shows the added efficiencies as technical, economical and environmental individual efficiencies.

Figure 20 depicts the calculated individual efficiencies in three dimensions as 3D Portfolio. Concerning the examined oil filter casings, the graphic charts depicts what Table 5 expresses in numbers:

Technically the easy-change steel oil filter casing with the disposable aluminium has an advantage over the two other designs (nylon and aluminium diecast version). As Table 5 shows, the main reasons for that are the utilisation of the part (function safety) and the material characteristics.

Table 5: Summary of the single efficiencies and determination of the Life Cycle Efficiency (GN) through addition of the single efficiencies for three oil filter cartridge designs

Varianten →		Nutzwert 1: PA 6.6	Nutzwert 2: St	Nutzwert 3: Al
technische Kriterien	Werkstoffeigenschaften	0,385	0,485	0,440
	Verarbeitbarkeit	0,739	0,695	0,627
	Konstruktion	0,380	0,355	0,400
	Montage	1,290	1,020	1,170
	Bauteil-Nutzung	1,071	1,545	1,385
	Σ technischer Nutzwert (TN)	3,865	4,100	4,02
wirtschaftl. Kriterien	Externe wirtschaftliche Kriterien	1,240	2,170	1,890
	Interne wirtschaftliche Kriterien	1,080	1,020	1,160
	Σ wirtschaftlicher Nutzwert (WN)	2,320	3,19	3,05
umweltliche Kriterien	Werkstoffherstellung	0,165	0,460	0,280
	Bauteilverarbeitung	0,165	0,470	0,280
	Bauteil-Nutzung	1,950	2,893	1,366
	Recycling	0,320	0,460	0,450
	Entsorgung (Deponie)	0,150	0,250	0,250
	Σ umweltlicher Nutzwert (UN)	2,75	4,53	2,63
	Ganzheitlicher Nutzwert Σ TN + WN + UN = GN	8,935	11,82	9,7

Esconomically the steel version has clear advantages. In particular the marketability and image in the market speak in the favour of steel.

Environmentally the disposable easy-change steel oil filter boasts the greatest advantages as compared to the disposable polyamide and aluminium diecast casings. The main reason for this is the low weight of the part and the reutilization, respectively disposal.

Figure 20 summarises the results of Table 5. Hereby one should always be aware of the fact that the Life Cycle Analysis and the subsequent evaluation is based on today's technology. The technical and economical advantages of an easy-change steel filter are based on decades of experience in mass production, while the polyamide design is still in the stage of being a prototype. This elaboration might look very different in 5 years from now, which again would underline the snapshot character of such analysis. Certainly the environmental result is astonishing. After qualitative consideration one would hardly have selected the easy-change oil filter as a disposable product. The example oil filter casings thus proves how important Life Cycle Analyses are for the decision-making process concerning product development.

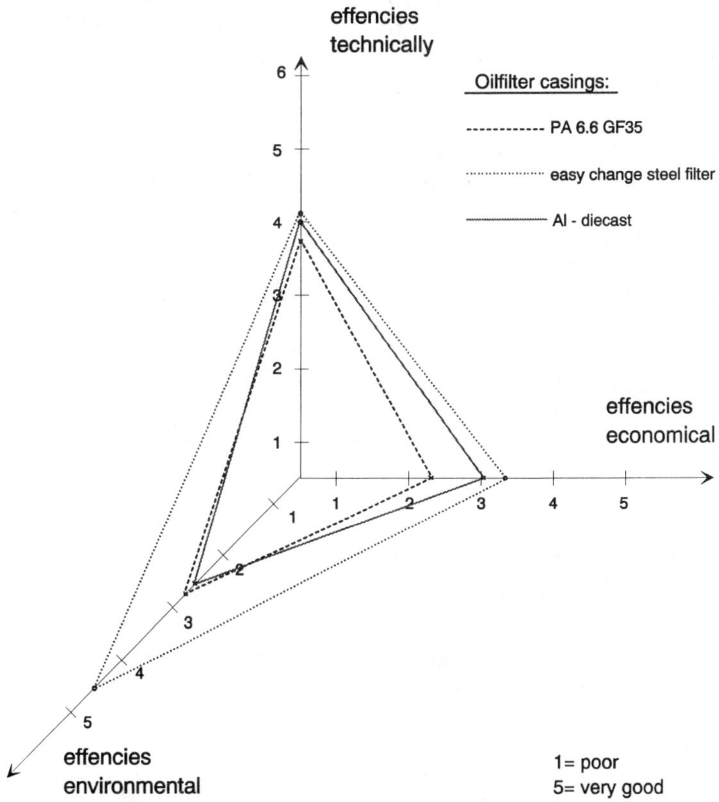

Figure 20: 3D Portfolio presentation: Life Cycle Efficiency as a triangle surface on the example of oil filter casings made of steel, aluminium diecast and polyamide 6.6 GF 35

When wishing to integrate the costs into this evaluation system, the Life Cycle Efficiency-Cost quotient (GNKQ) can be calculated (Figure 18)

$$GNKQ = \frac{GN}{production\ nand\ assembly\ costs}(\frac{1}{DM})$$

The calculation of the GN can be done by adding the three single efficiencies (Table 5), or the surface gravity centres or the triangles drawn in Figure 20 can be used for that purpose. By assuming the production and assembly costs, the numbers determined in Table 6 emerge, and a preference emerges from these examples.

Final conclusions on the evaluation

The selected degrees of performance play an important rule in the determination of the single efficiencies. They should therefore be selected in a team, if possible, while taking the special situation of the individual case to be evaluated into consideration. Besides the equal addition, any other weighting can be introduced for the determination of the Life Cycle Efficiency (technical efficiency, economical efficiency, environmental efficiency). The same method of the Efficiency Cost-Analysis is also to be used again.

The following considerations on limiting values may be useful for the evaluation of the Life Cycle Efficiency (maximum 5) divided by low production and assembly costs would be the ideal case of an evaluation and therefore a low (good) ranking. Thus the real cost dominance will also reflect in the evaluations. But other emphases as desired can also be placed depending on each individual case. Shall, for instance, a technical characteristic be emphasised, high percentages are given as a rule and the desired weighting will be achieved.

Finally it must be emphasised that the offered Life Cycle Evaluation Method as demonstrated on the parts of oil filter casings can be, but does not have to be, an aid.

Table 6: Cost-related result of the Life Cycle Efficiency-Cost-Analysis (GNKA) with ranking for three oil filter cartridge designs

Datum	Variante Nr.	Beschreibung **Ölfiltergehäuse**		Ganzheitl. Nutzwert GN	Σ Herstell-kosten [DM]	$\frac{GN}{HK}\left[\frac{1}{DM}\right]$	Rang-folge
	1	**PA 6.6 GF 35**		**8,94**	**50,-**	**0,18**	**2**
	2	**Stahl Wechselfiltergehäuse**		**11,82**	**48,-**	**0,25**	**1**
	3	**Al - Druckguß**		**9,70**	**70,-**	**0,14**	**3**

The authors want to offer an instrument for a better management of the flood of data, but they do not want to create a dogma with legislative power. The supplier of the oil filter, for instance, will want to minimise his production costs and offer environmental advantages besides good technical characteristics, while the automobile manufacturer, apart from a good purchase price, will especially weight his own assembly costs, a low weight of the part and an unproblematic disposal. The offered Life Cycle Evaluation Method does justice to these individual boundary conditions. The Life Cycle Evaluation Method borrows from the well-known Efficiency-Cost-Analysis. It offers few answers to questions concerning the standardised evaluation of the toxicity of different gases, wastes, etc.

Through standardised values such as Maximum Work Environment (MAK) or Recommended Technical Concentrations (TRK) one can carry out weightings and evaluations of emissions. However, as long as many important questions concerning the atmospheric chemistry remain unanswered, a generally mandatory evaluation method will not be possible. The Life Cycle Evaluation Method is a tool for the free enterprise economy. It helps its users answer each existing individual situation from their own points of view.

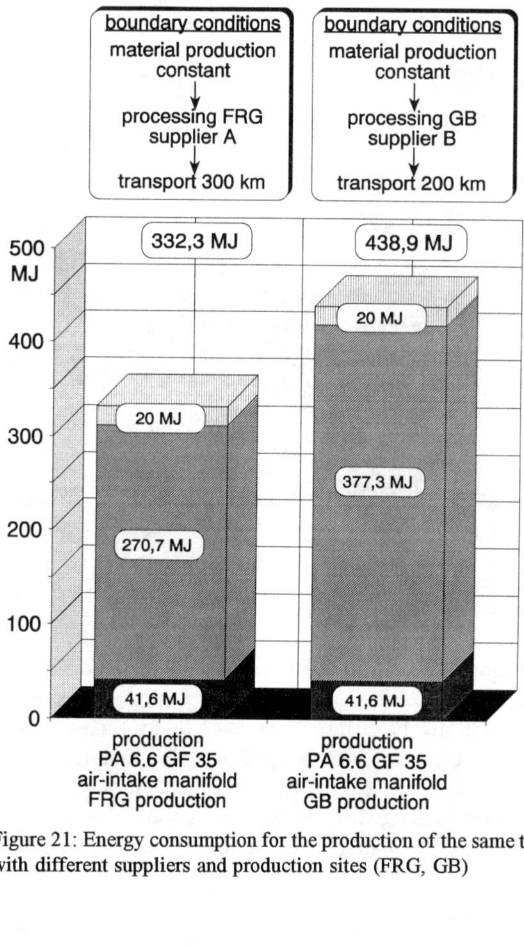

Figure 21: Energy consumption for the production of the same type of air-intake manifolds (PA 6.6 GF 35), but with different suppliers and production sites (FRG, GB)

Final conclusions and prospects regarding the Life Cycle Analysis of oil filter casings

For the purpose of this Life Cycle Analysis, it was attempted to make use of an oil filter design that is as representative as possible. However, it certainly seems possible that larger filter versions could lead to different results. There especially, however, weight savings of the aluminium and synthetics versions versus steel could become apparent. Furthermore the data basis for PA 6.6 has to be expanded considerably, firstly in the different synthesis directions.

These versions, however, can lead to the necessity to re-check the above calculation if the filter size is changed, but especially if other types of vehicles (such as diesel) with essentially different energy consumptions are examined. At the end of this analysis the reader will probably wish for a generalised statement in favour of or against one or the other material.

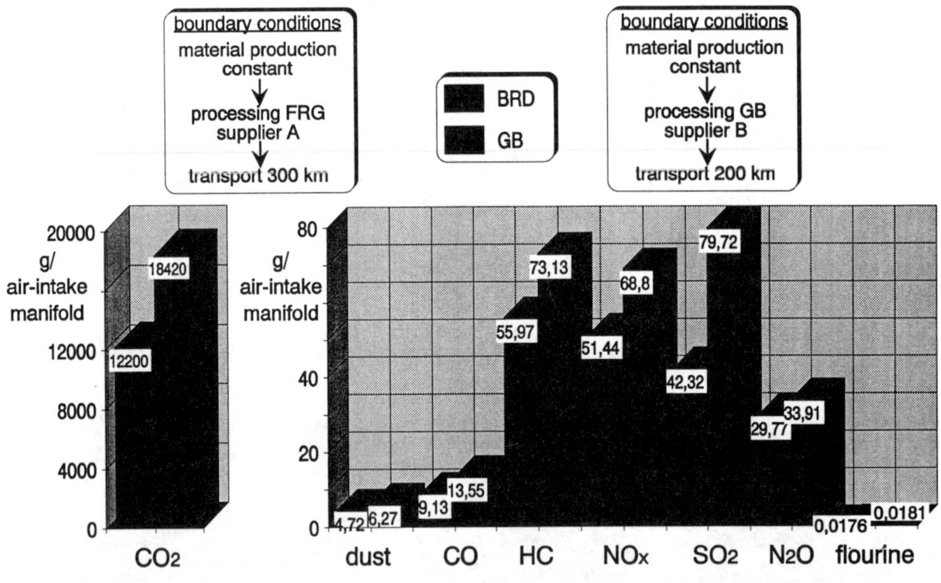

Figure 22: Atmospheric emissions during the production of the same type of air-intake manifolds (PA 6.6 GF 35), but with different suppliers and production sites (FRG, GB)

But, as has been emphasised again and again, we caution against doing exactly that. The statement made here apply only to the parts described above. Other constructions with different boundary conditions can lead to completely different results. Merely the selection of a different manufacturing country (such as England), causes considerably greater amount of waste. Also the energy mix greatly differs there as compared to the Federal Republic of Germany, which alters the energy and emission balance. Figures 21 and 22 show the different actual energy consumptions and emissions in the Federal Republic of Germany and in

England, as demonstrated on a air-intake manifold made of PA 6.6 GF 35 - and therefore easily transferable to a PA oil filter casing. Life Cycle Analyses are therefore also a tool for the evaluation of decision concerning the site.

When summarising all arguments, it remains to be said, that there is no way around a consideration for weak point analyses that differentiates between materials, methods and parts.

Life Cycle Analyses are snapshots and allow the preparation of declarative weak point analyses for certain constructions and supply situations. Generalisations are not permissible. After all, we do not make conclusions concerning the overall steel industry based on the balance of capital transactions of the John Doe Metal Processing Company for the year of 1993. Life Cycle Analyses are similarly useful as balances of capital transactions, as long as certain rules are kept. However, they can also be manipulated if the preparing person intends to do that.

The above inventory brings up many questions. How are emissions in the air or water and in comparison to dumping volume to be evaluated? Probably not in the near future, therefore the suggestion of a subjective method is to be evaluated. Are there objective rules for operative balances (evaluations) similar to those for inventories? A lot of action is required here!

The example demonstrated here shows today's status of Life Cycle Analyses. The manifold blanks, especially concerning operational balances, must jointly be filled in during the coming years.

Reference

1 Eyerer, P.; Dekorsy, Th.; Schuckert, M.: Ganzheitliche Bilanzierung von Bauteilen, Vortrag: CIS, Salzburg 9/1990

2 Dekorsy, Th.: Ganzheitliche Bilanzierung als Instrument zur bauteilspezifischen Werkstoff- und Verfahrensauswahl, Dissertation, Universität Stuttgart, 9.2.1993

3 Eyerer, P.; Schuckert, M.; Dekorsy, Th.; Pfleiderer, I.: Ganzheitliche Bilanzierung von PKW Ansaugrohren, Aluminium 69 (1993) 5, S. 435-440

4 Schuckert, M.; Dekorsy, Th.; Pfleiderer, I.; Eyerer, P.: Developing a comprehensive balance of an automotive air intake manifold; German Plastics 83 (1993) 3, S. 16-19

5 Schuckert, M.; Dekorsy, Th.; Pfleiderer, I.; Eyerer, P.: Methodik zur Ganzheitlichen Bilanzierung, unveröffentlichter Forschungsbericht der PE Product Engineering GmbH, Ostfildern 1992

6 Kindler, H.; Nikles, A.: Energieaufwand zur Herstellung von Werkstoffen Kunststoffe 70 (1980)

7 Birnbaum, U.; Wagner, H. J.: Einheitliche Berechnungen von CO2-Emissionen. Energiewirtschaftliche Tagesfragen 42 (1991) 1, S. 78-80

8 Pfleiderer, I.: Das Software-System zur Ganzheitlichen Bilanzierung. Vortrag: Fachkongreß zur Ganzheitlichen Bilanzierung, Bad Nauheim, 15./16.6.1992

9 Pfleiderer, I.: Das Software-System for LCA. Vortrag: LCA Software Workshop der UETP-EEE in Gothenburg, Sweden, 20./21. August 1992

10 Ahbe, Stefan; Braunschweig, Arthur; Müller-Wenk, Ruedi: Methodik für Ökobilanzen auf der Basis der ökologischen Optimierung, Schriftenreihe Umwelt Nr. 139/90, Bern
11 Suter, P.; Hofstetter, P.: Die ökologische Rückzahldauer; Schweizer Ingenieur und Architekt Nr. 491/89, S. 1342-1346
12 Mc Kinsey & Company: Integrated substance chain management, Association of the Dutch Chemical Industrie, 1991
13 Swedish Environmental Research Institute: A system for calculating environmental impact, 1991
14 Warnecke, H. J.; Bullinger, H.-J.; Hichert, R.; Voegele, A.: Wirtschaftlichkeitsrechnung für Ingenieure, Hanser-Verlag München, 1990, S. 152f

Ecological and energy balance with regard to the various materials used in the automobile industry

Siegfried Schäper and Hans-Günther Haldenwanger

Whereas not so long ago products won awards for outstanding technology alone, ecological characteristics are now playing an increasingly prominent part. And it is not so much the fulfilment of statutory requirements as the voluntary provision of extras and the anticipation of technical standards and legal deadlines that is required.

Prospective customers' emotionally coloured views on such points as "outlawing" PVC or tropical hardwoods can scarcely be ignored by marketing strategists. As such things are a matter of fashion, however, they cannot be a proper basis for strategic decisions. These require objective methods of assessment with long-term reliability. In the political forum, too, the increasingly public debate on technical systems and their consequences would be much improved by the application of such objective, reliable criteria rather than agitation and emotion.

It is expected of an ecological balance that it should provide and objective yardstick of this kind.

Ecological balance - energy balance

As defined by the German Federal Department of the Environment [1], the ecological balance is "as comprehensive a comparison as possible of the effects on the environment of two or more different products, product groups, systems, processes or procedures. It serves to highlight weaknesses, to improve the ecological characteristics of the products, to aid decision making in procurement and purchasing, to promote environment-friendly products and processes, to facilitate comparison of alternative procedures and to produce recommendations for action. Depending on the nature of the questions involved, this comparison is expanded to include further aspects such as an assessment of the efficient use of funds in environmental protection."

Accordingly, measures taken to improve environment-friendliness must be viewed holistically. The corollary of this includes at the least the following two points.

- Firstly, the entire product cycle from the acquisition of materials and manufacture to the use of the product and disposal after use in as environment-friendly a way as possible, for instance by recycling for re-use in a further product cycle, must be taken into consideration.
- Secondly, at least the major environment-relevant effects, namely water, soil and air pollution, together with the consumption of scarce resources must be taken into account.

In this way relationships and effects arising between different stages of the product cycle, between various kinds of harmful emissions and different ways in which scarce resources are consumed can be clearly recognized and, if necessary, taken into account. The aim must be to define an optimum problem solving strategy. This optimum will be a compromise which avoids the one-sided overemphasis of a single feature that can result from linear thinking.

A comparative, quantifying approach to assessment does, of course, presuppose that, on the one hand, the environment-relevant effects of the overall product cycle must be considered as completely as possible [2] and, on the other, that even basically incompatible features can be drawn into the same equation. One inquiry might refer, for instance, to which of the following technical options would be preferable.

- The one with a lower CO_2 emission?
- The one that runs more quietly?
- The one that causes less noise at the workplace?
- The one that will take up least space when disposed of in a landfill?

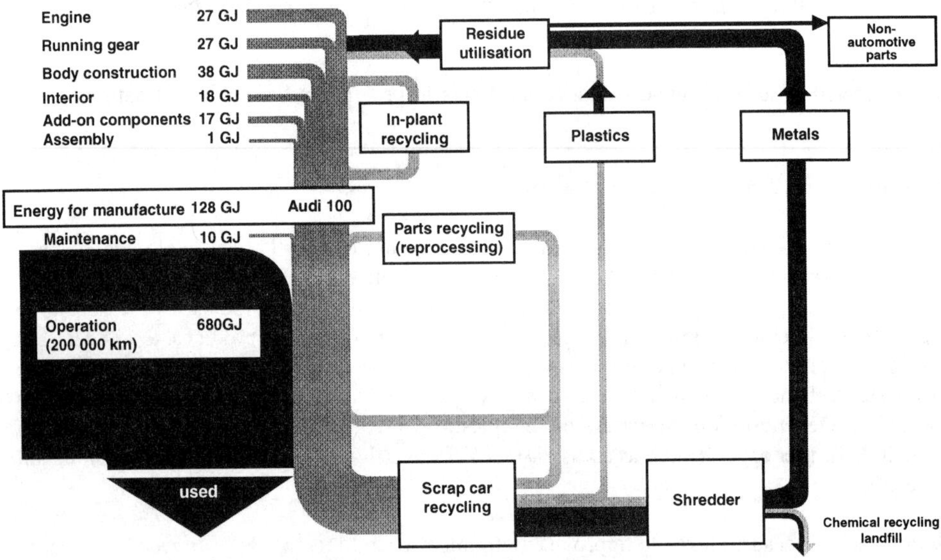

Figure 1 Use of energy for a mid-size private car in conventional steel construction according to Schaefer/Woyke[1]

Judgements and conclusions drawn from ecological balance sheets are necessarily influenced by political priorities. Waste water containing NaCl will be a problem on the upper reaches of a river - but not if it is discharged into the same river from a location on its estuary. The

[1] Woyke, O.: Kumulierter Energieaufwand zur Herstellung eines Mittelklasse-PKWs am Beispiel eines Audi 80 (Cumulative expenditure of energy involved in the production of a medium range car, taking the Audi 80 as an example) graduate diploma thesis at Lehrstuhl für Energiewirtschaft und Kraftwerkstechnik Technische Universität München, 1992

statements made in ecological balance sheets may be objective but they are basically dependent on particulars of time and place and should not be taken as having general validity [3].

Some of what is expected of an ecological balance sheet remains unfulfilled for other reasons, eg. because methods are not standardized or the practice of global sourcing requires statements of comparable validity world-wide or because of the impossibility of applying German or European criteria world-wide. In addition to this, the expenditure of time, money and effort involved in setting up a full ecological balance sheet for a complex product is considerable and poses the question of whether the needs of the interested parties could not be met in some other way.

As another approach to the problem the use of an **energy balance sheet instead of the ecological balance sheet** ought to be discussed. Several good arguments in favour of this are gone into below with the aid of Figures 1 and 2.

Figure 1 shows the use of resources in units of energy for a mid-size car of conventional steel construction[1]. Of the total energy consumed in the life of the vehicle 83% is accounted for by fuel. This breaks down as follows: 127.3 GJ for production (including acquisition of the raw materials, transport, handling and assembly), 10 GJ for maintenance, 681 GJ for actually driving the car, given average petrol consumption of 9.4 litres per 100 km for a distance of 200.000 km, and 2 GJ for scrapping the vehicle (1.5 MJ/kg for dismantling and shredding[2])

The total energy requirement came to 820 GJ. With a vehicle service life of 15 years (fully galvanized body) a total distance driven of 200,000 km amounts to 13,333 km per year, which is average use. If actual driving accounts for the lion's share of the energy balance (83%) the figure for the - at least as far as quantity is concerned - most significant undesirable substance produced, namely CO_2, is as high as 93%, equal to approx. 49 tonnes produced by actual driving of a total of approx. 53 tonnes![2]

Figure 2 shows the flow of energy in the Federal Republic of Germany in the year 1989. The imported or domestically produced fossil or other fuels in the form of coal, oil, natural gas and nuclear energy together with a relatively small proportion of regenerable energy are mainly converted into end products in large-scale industrial plants such as refineries, power stations, district heating power stations and coking plants. Figure 2 shows district heating, coal, electricity, oil and gas as the most important end products for consumption in industry, transport, households, public authorities etc. and for export and stockpiling. It draws attention to the following important details.

- 85% of the Federal Republic of Germany's energy supply comes from fossil resources.

[2]Author's Figures derived from the IKP project "Holistic balances"
Eyerer, P. at al.: Ganzheitliche Bilanzierung von Automobilbauteilen, (Holistic balance sheets for automobile parts), Joint research project of the Institute für Kunststoffprüfung und Kunststoffkunde (IKP) Universität Stuttgart and companies associations of the steel, aluminium, plastics and car industries, 1991 to 1993.

- The entire system is principally geared to the production of energy, even though the input "fossil resources" is in fact ambivalent: they can be used both as a source of energy and as raw materials. A total of only 6.3% is used as raw material, equal to 10% of what reaches the end user.
- 57% of energy resources are lost during conversion into end products and as energy consumed by end users.
- Of the energy put at the disposal of end users 25% is in the form of fuel for transport where it is exploited with the lowest efficiency of all the sectors of consumption, ie. a mere 18%!

In a situation like this the use of energy always entails consumption of resources and pollution of the environment.

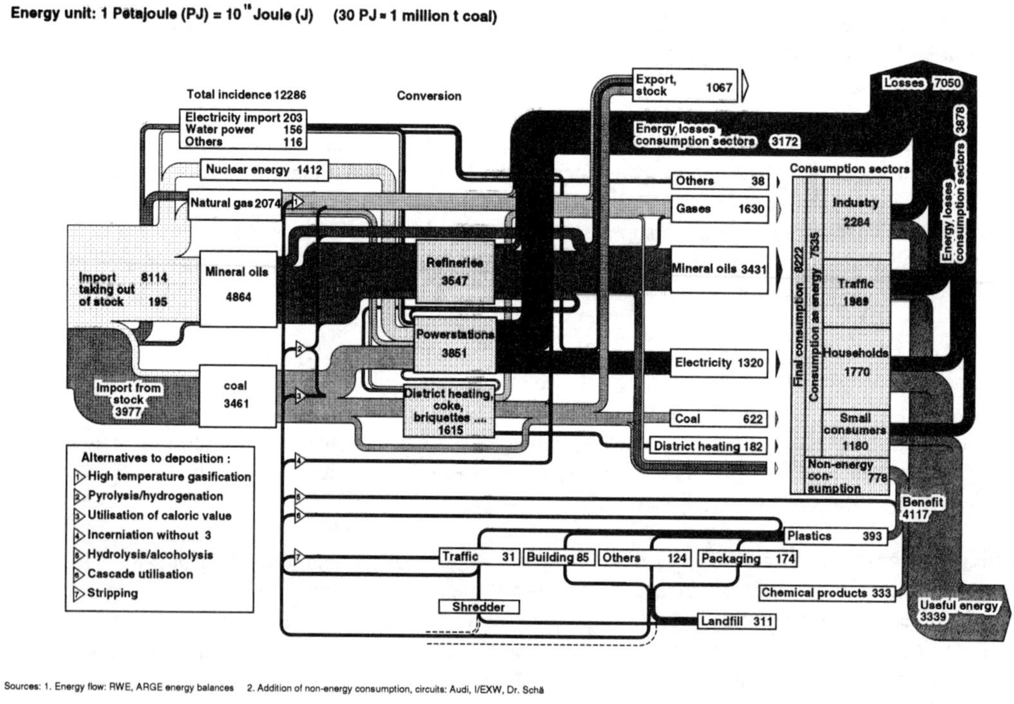

Figure 2 Energy flow in the Federal Republic of Germany, 1989 (not includet former GDR)

When ecological balance sheets are discussed in the car industry the main issue is mostly the choice of materials. It must be presumed the effects on the environment of the production and use of alternative materials is of equal importance and that the energy consumption involved in this is a notable factor in their ecological suitability. The energy balance sheet, being the essential part of the ecological balance sheet, accordingly has the same consequences for car production as the much more cumbersome and problematic ecological balance sheet. The energy balance sheet also has the advantage of being more generally applicable and not so

dependent on political priorities. This does not mean that the pollution hazards dependent on the materials used or the consumption of scarce resources are to be uncritically accepted - this issue is frequently regulated by law. Accordingly, there seems to be no doubt about the viability of using the energy balance sheet as a basis for a development strategy.
Examples of energy balance sheets are accordingly discussed below, the influence of weight on fuel consumption being dealt with first.

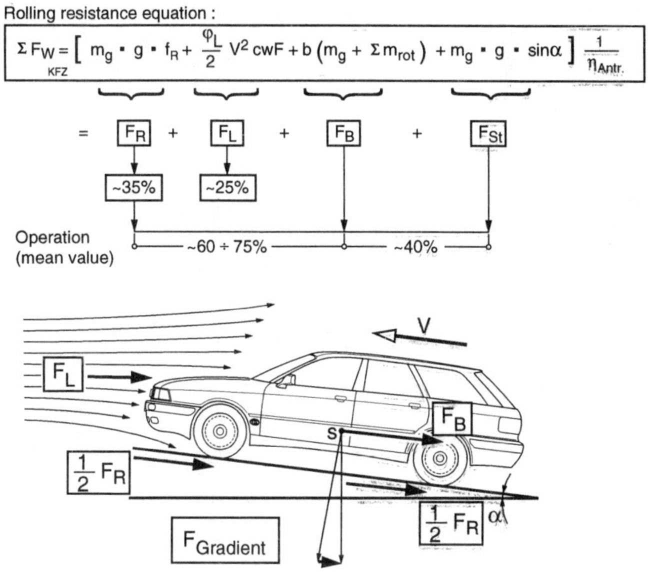

Figure 3 Rolling resistance equation

Figure 3 gives the equation showing the forces of resistance acting on the vehicle:

$$\Sigma \mathbf{F}_W = \mathbf{F}_L + \mathbf{F}_R + \mathbf{F}_B + \mathbf{F}_S$$

The terms air resistance $\mathbf{F}_L$, road drag and friction $\mathbf{F}_R$, acceleration resistance $\mathbf{F}_B$, and resistance due to gradient $\mathbf{F}_S$ include, with the exception of the air resistance which accounts for approx. 25% of resistance in the usual test cycles, the car´s weight three times and the rotating masses once. Fuel consumption is, in practice, substantially dependent on weight. ADAC test Figures show that on an average of all cars tested weight accounts for approx. 0.7 l/(100 km · 100 kg) (Figure 4 [4])

The weight factor is less important in the case of vehicles with a low specific performance: over 18 kg/kW, eg. 0.35 l/(100 km · 100 kg). This is higher for vehicles with a high specific performance: up to 11 kg/kW, eg. 0.85 l/(100 km · 100 kg). It is also dependent on the test cycle used. In the case of the Audi 80 the Figures for a third mix in the EG cycle are 0.6 l/(100 km · 100 kg). This is in accord with market observation and therefore corresponds to realistic conditions. Fuel economies are related not only to the calorific value of the fuel

(32.22 MJ/l) but also to the energy expended in supplying it (4 MJ/l). As in Figure 1, an average vehicle life of 200,000 km was assumed. In the case of company cars in the top class and upper mid-size the mileage will be higher, which lends support to the arguments in favour of cutting down on superfluous weight.

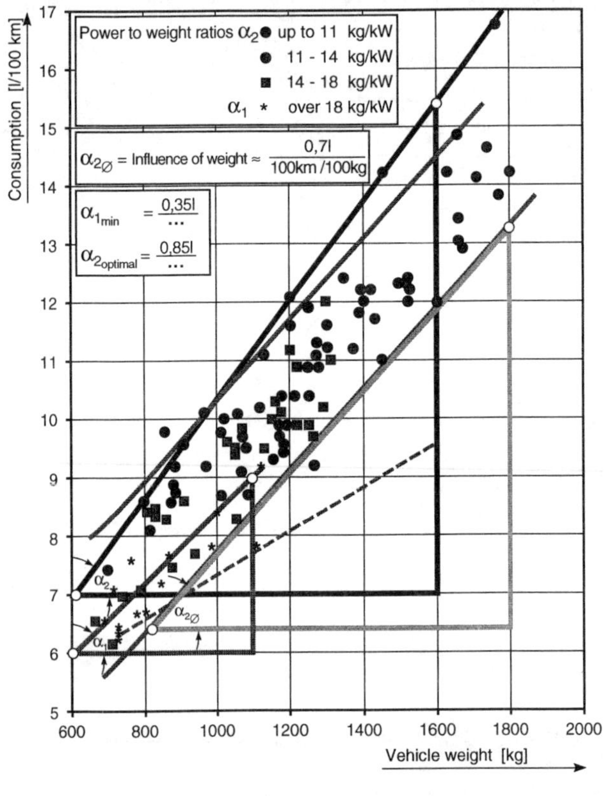

Figure 4 Actual fuel consumption according to ADAC [4]

Description of the service life cycle of various components, taking the example of a valve bonnet

The valve bonnet is an example of a component that can function when made of a great variety of materials. Accordingly, it serves as a good model for an investigation into the environmental suitability of various versions and the differences between them. *Dangelat*[3] and

[3]Dangelat, M: Ressourcenschonung durch Stoffrecycling unter Voraussetzung von positiver Energiebilanz am Beispiel Zylinderkopfhaube (Saving resources by recycling on the basis of a positive energy balance sheet, taking the example of a cylinder head cover) graduate diploma at Fachbereich Fahrzeugtechnik und Konstruktion TH Zwickau, 1992

Huber[4] have pursued these questions; the following figures are taken from their work, or, as in the case of the alternative materials aluminium sheet, die cast magnesium and titanium, derived from it.

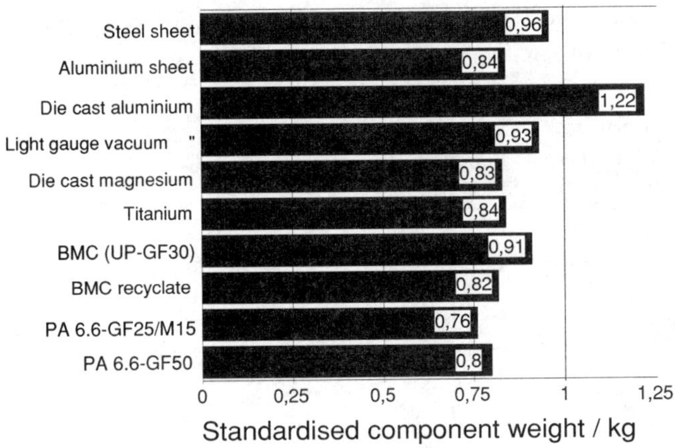

Figure 5 Standardised weights of valve bonnets in various materials

The weights of the components made in the various materials are given in Figure 5, the real variants for 3, 4 and 5 cylinders with 2 and 4 valves having been converted to match the conditions of a 4 cylinder 2 valve model with a rubber/metal gasket/screw thread fastening and seal system. The differences in weight achieved using lightweight materials as opposed to the usual Audi technology - sheet steel in the 4 cylinder 2 valve model - are relatively small: the biggest difference (20%) can be attained with PA 6.6-GF25/M15; sheet aluminium, die cast magnesium, titanium, BMC recyclate and PA 6.6-GF50 achieve 12.5% and 16.5%; new BMC reaches 5%; vacuum die cast thin aluminium is only 3% lighter. The version in ordinary die cast aluminium is actually 27% heavier than the latest sheet steel model! The perhaps surprising advantage of BMC recyclate over the new material is due to the fact that the ERCOM recycling technique replaces part of the mineral filler with Polyester resin, reducing the density of the melt from 1.95 g/cm^3 to 1.75 g/cm^3. This 10% density advantage has also been ascertained on the SMC front end of the VW Passat. Apart from anything else, the scope for reducing the weight of a component like the valve bonnet are very limited owing to the mechanical stress arising from the seals and fastenings and the minimum wall thicknesses imposed by die cast technology.

The use of energy involved in production and operation of the valve bonnets made of various materials and also in that of the valve bonnets referred to in Figure 5 is depicted in Figure 6. As far as production is concerned, by far the most favourable version is the sheet steel one. The next most favourable version, made of BMC, requires 3.2 times as much energy and the heaviest, least favourable version in die cast aluminium needs 12 times as much. Fuel

[4]Huber, K.: Titanium materials and lightweight structures; trainee thesis 1992 l/(100 km · 100 kg)

consumption is in proportion to weight (Figure 5), which means that the PA 6.6-GF25/M15 version is 20% better than the steel one. This is also the best overall, requiring 6% less than the sheet steel valve bonnet. The second PA valve bonnet, weighing 17% less than the steel one, has roughly the same energy requirement. The 13% lighter sheet aluminium valve bonnet fails to match up to the steel version: its overall energy needs for production and operation are 18% higher and its requirement for production is six times as high.

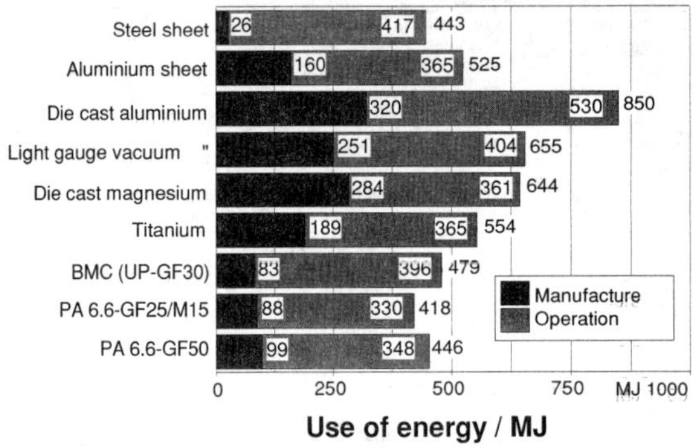

Figure 6 Use of energy for the manufacture and operation of valve bonnets in various materials (200,000 km)

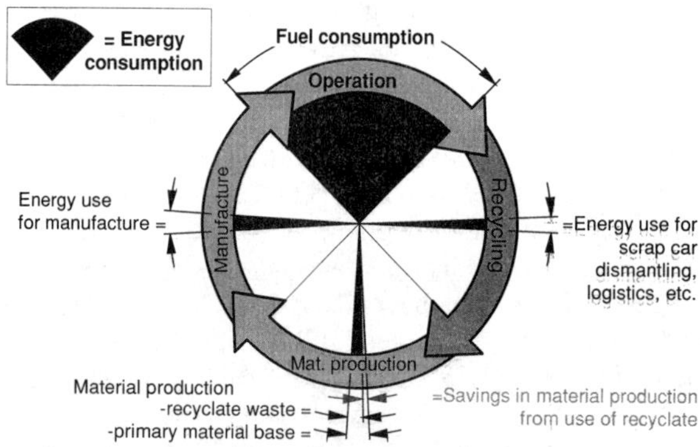

Figure 7 Schematic of the total energy balance for a material circuit

Figure 7 shows the principle of an overall product cycle comprising provision of materials, production, use and recycling. Energy must be expended in all stages of the cycle. It is, however, a fact that the use of recycled materials for the valve bonnets involves substantially less energy than when primary raw materials are used. The saving is so great that the initial outlay for recycling is clearly recovered.

Figure 8 shows the energy expenditure involved in the production of valve bonnets from primary raw materials (identical with Figure 6) compared to that required when using recycled materials. The assumption was made that the recycled materials were equal in performance to new material and could therefore be used unmixed. Having said this, it must be added that 40% of the BMC recyclate, ie. the proportion of filler, was replaced by ERCOM recyclate. The components weigh the same as those made of new material - with the exception, once again, of the BMC component which is 10% lighter. This has a corresponding effect on fuel consumption (for this reason SMC recyclate beats the sheet steel version on the overall energy requirement for production and operation). Using recycled material the steel valve bonnet is 35% cheaper to produce than with new material. As, except in the case of BMC, the savings achieved as a result of recycling are higher than with steel, namely between 55% (magnesium) and 77% (PA 6.6 GF 50), the scale of low-energy materials is adjusted accordingly. Not only PA 6.6 GF 50, BMC and sheet aluminium but even titanium does better than steel. Owing to the high minimum wall thicknesses and the resultant high weight of the component the die cast aluminium and magnesium versions remain less advantageous. Figure 8 provides an impressive illustration of how wrong it can be to judge the energy required for a particular component simply on the basis of the amount needed for production - a mistake that is constantly being made in popular publications. The overall expenditure of energy required by the lightest valve bonnet, the PA 6.6 GF 25/M15, made from new material is 340% of that needed by the sheet steel version (which has the lowest energy needs for production). Using recycled material for both versions production of the polyamide version still requires 130%. However, in both cases the overall energy requirement covering production and operation puts the 20% lighter component clearly ahead. Its advantage over the steel component is 6% (new material) and 19% (recyclate). Despite a recycling saving of 62% die cast aluminium remains the least favourable of the versions examined, owing to its weight.

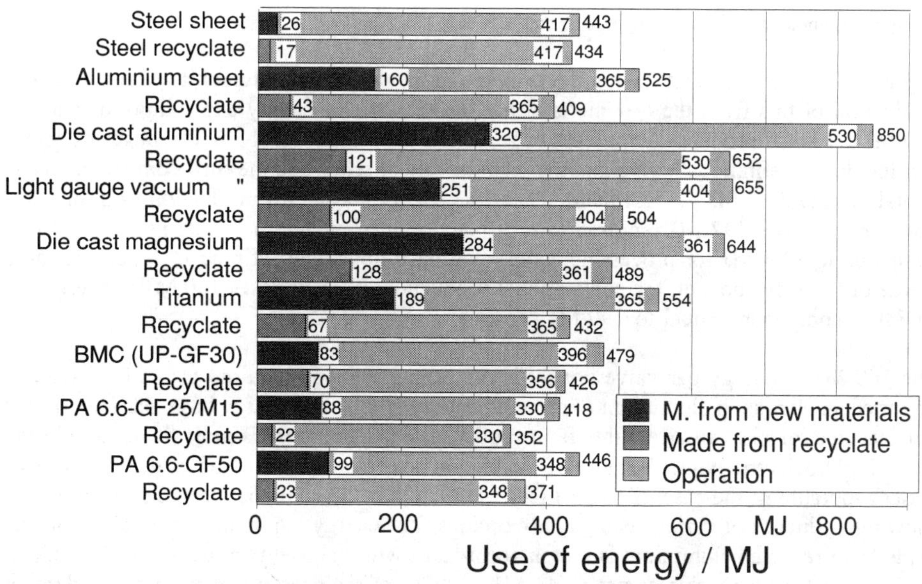

Figure 8 Use of energy for the manufacture and operation of valve bonnets in primary and secondary materials (200,000 km)

The amount of electricity required in the production of the light metals aluminium, magnesium and titanium from raw materials is relatively high. Recycling, which cuts this energy requirement and improves the entire energy balance sheet, is a particularly attractive option in the case of light metals.

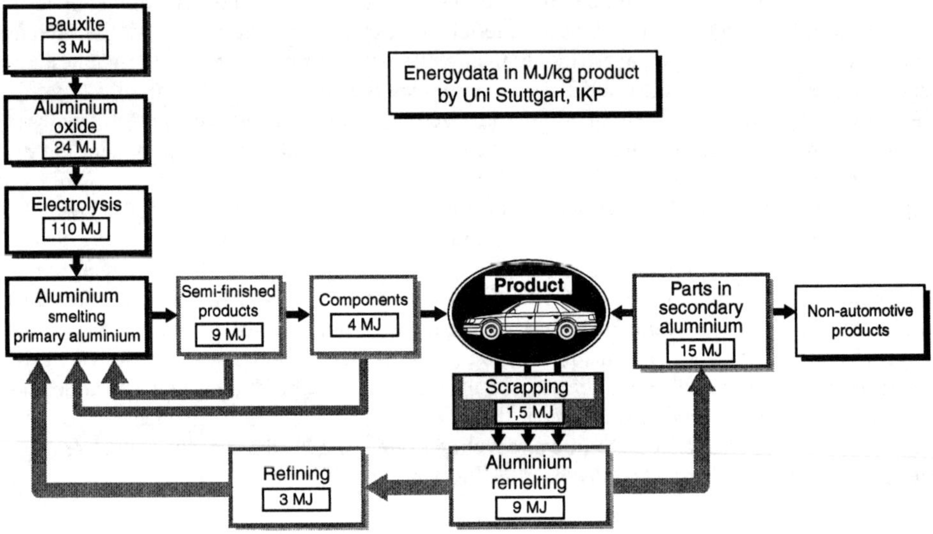

Figure 9 Aluminium manufacturing process and recycling

Figure 9 illustrates this fact, taking the example of an aluminium automobile wing[2]. Whereas production of this from the raw material bauxite requires an energy expenditure of approx. 150 MJ per kg of product, the process of recycling aluminium involving the stages of scrap, melting down, refining, alloying and the manufacture of semis and the final component needs a total of only 26.5 MJ, ie. less than 20% of the energy needed for production from primary raw material. The 137 MJ required to turn bauxite into smelter shop aluminium has to be compared to 13.5 MJ for processing scrap, melting it down and refining it. Much the same is true of the valve bonnet, the conventional production of which takes 160 MJ of energy for an 840 g component, equal to 190 MJ per kg.

The 160 MJ of energy per valve bonnet is the primary energy required to produce average imported smelter shop aluminium ("imported energy mix"). The energy base, particularly for igneous electrolysis, can affect this figure substantially. The figure is most favourable in the case of hydroelectric power because of the high efficiency of this method of generation; it is least favourable in the case of fossil fuel power stations. The 5 upper sections of Figure 10 show the influence of various energy base options. The energy requirement for valve bonnets made from recycled aluminium is met in accordance with the energy structure of the Federal Republic of Germany: this comes to 43 MJ, ie. 27% of the energy expended on production from raw materials on the basis of the "imported energy mix". 14% of the "imported energy mix" figure could be saved using hydroelectric power. The average "energy mix of the western world" figure is 8% higher than the "imported energy mix". Using energy from fossil fuels only 40% more would have to be expended. Owing to the unchanged weight of the component (840 g) the fuel requirement remains 365 MJ. The 72% saving on energy in production owing to recycling results in a saving in the overall balance sheet of 22%. On the other hand, the effects of the primary energy structure disappear, ie. of the 14% gain from the use of hydroelectric power 4% remains and is matched by a 2% loss in the energy mix of the western world. Of the 42 % disadvantage owing to the sole use of energy produced from fossil fuel 13% remains in the overall balance. In Figure 6 the "imported energy mix" put the sheet aluminium valve bonnet in fifth place. Changing the energy structure would have altered nothing here, except in a worst case: using fossil fuel energy the sheet aluminium version would have more extravagant than the titanium version.[5]

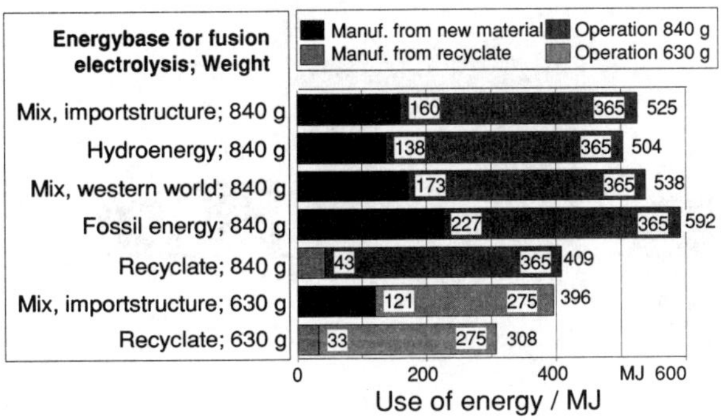

Figure 10 Use of energy for the manufacture and operation of valve bonnets in sheet aluminium, influence of the energy base of Al electrolysis (200,000 km)

Far more important than the issue of the energy structure on which bauxite mining, aluminium oxide production and electrolysis are based is that of cutting weight to a minimum. The difference between sheet steel and sheet aluminium was only 12.5% because substantial parts of the valve bonnet had to be solidly dimensioned owing to the principle of the fastening and seal system adopted. If this could have been departed from and the design could have relied on bending resistance - eg. fastening with a central brace and stick-sealing instead of rubber/metal gaskets - which would have brought the weight down to 630 g (Figure 10, bottom), ie. 35% less than the sheet steel version (960 g), an aluminium valve bonnet made of new raw material would have come out more favourably in the overall energy balance than the 840 g component made of recycled material. In Figures 6 and 8, too, the 630 g component would have beaten all the others by a clear margin. Using recycled material only the overall balances of the polyamide versions would have come off better. In this case minimizing weight would be more important ecologically than considerations of energy structure or raw materials versus recycled materials.

[5]Titanium valve bonnet on the basis of the "western world energy mix".

Description of the useful life cycle of two upper mid-size cars, one with a conventional body and the other with the Audi Space Frame

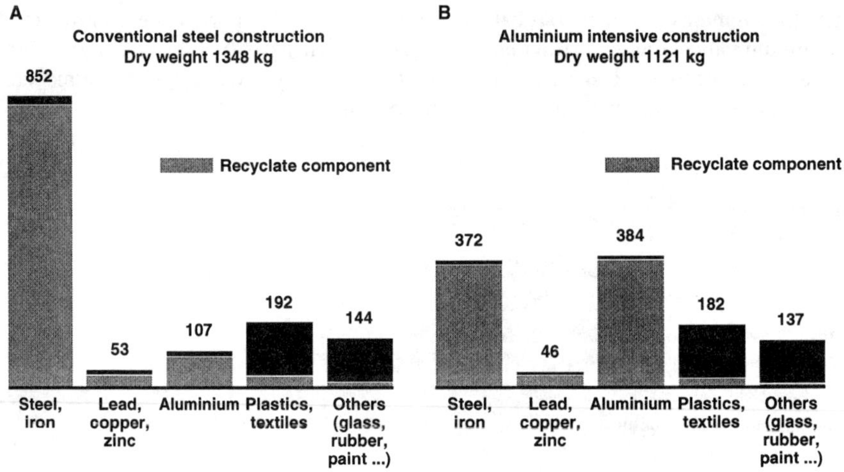

Figure 11 Material components of a mid-size private car (based on Audi 100/V6)

The so-called dry weight of an Audi 100 (front wheel drive, V 6 engine and manual gear shift) is 1348 kg. The Audi Space Frame version, making lavish use of aluminium, weighs 1121 kg. The essential difference is in the amount of steel used, this being reduced from 852 kg to 372 kg, and the increase of aluminium from 107 kg to 384 kg (Figure 11). In the Space Frame design the aluminium is used in the form of sheet, extruded sections and cast gussets (Figure 12 and 13).

Although the various alloys used differ they can be successfully recycled and put to their original uses again - a considerable advantage in the near future when manufacturers will be obliged to accept return of scrap cars for disposal. Audi have in fact been recycling special aluminium alloys in this way - an example is the production waste from door units - since 1983. The various alloys are kept separate for recycling so that, to a very considerable extent, sheet is re-used as sheet, extruded sections as extruded sections and casting or moulding alloy as casting or moulding alloy.

Laser assisted materials spectroscopy provides suitable techniques for the fully automated identification and separation of alloys by their components. Metallgesellschaft are conducting trials on the prototype of a laser detection system.

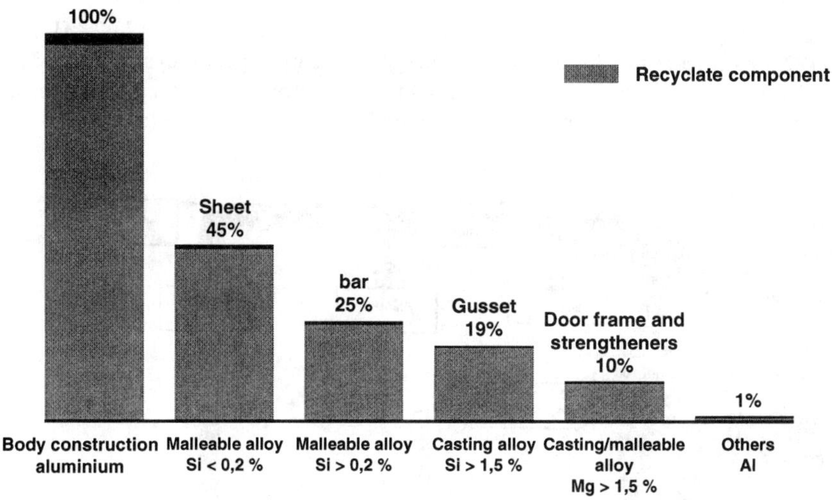

Figure 12 Aluminium components of the Audi Space Frame body

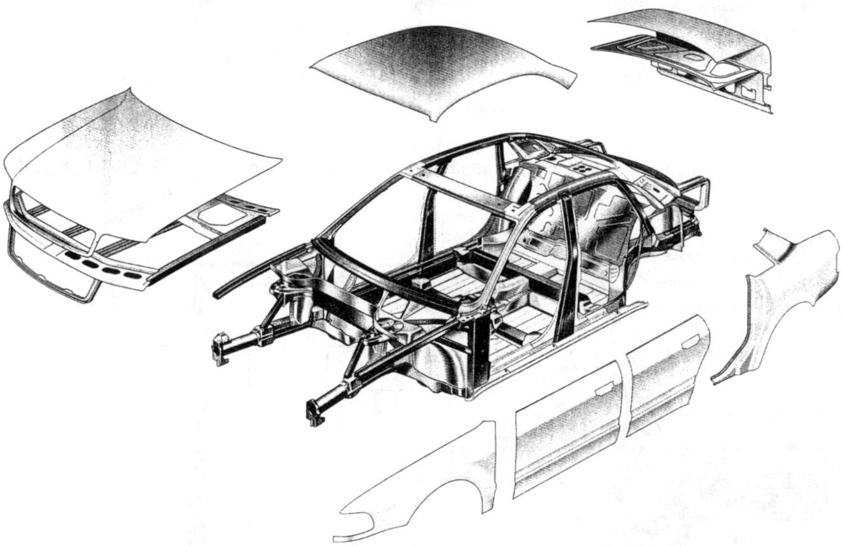

Figure 13 Audi Space Frame body

Figure 14 shows the energy flow for the production, maintenance and operation of a Space Frame vehicle made of new aluminium[6]. Figure 15 shows the same for recycled aluminium.

[6]Kawollek, R.: Ermittlung des kumulierten Energieaufwandes einer aluminiumintensiven Karosserie (Ascertainment of the cumulative expenditure of energy involved in a high-aluminium content car body) graduate diploma thesis at Lehrstuhl für Energiewirtschaft und Kraftwerkstechnik, Technische Universität München 1993

The energy requirements for production and maintenance are in the one case 173.2 MJ and in the other 122.5 MJ, equal to 23% and 17% of the total energy expenditures of 764.2 MJ and 713.5 MJ for a distance of 200,000 km. As in the case of conventionally designed vehicles, the fuel requirement is the main factor in overall energy expenditure (see Figure 1).

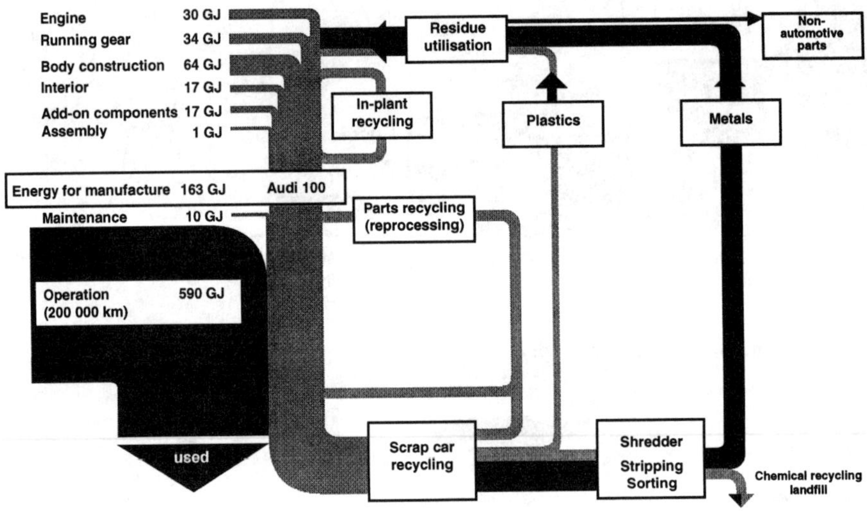

Figure 14 Use of energy for a mid-size car in aluminium intensive construction according to Schaefer/Kawollek (primary aluminium)

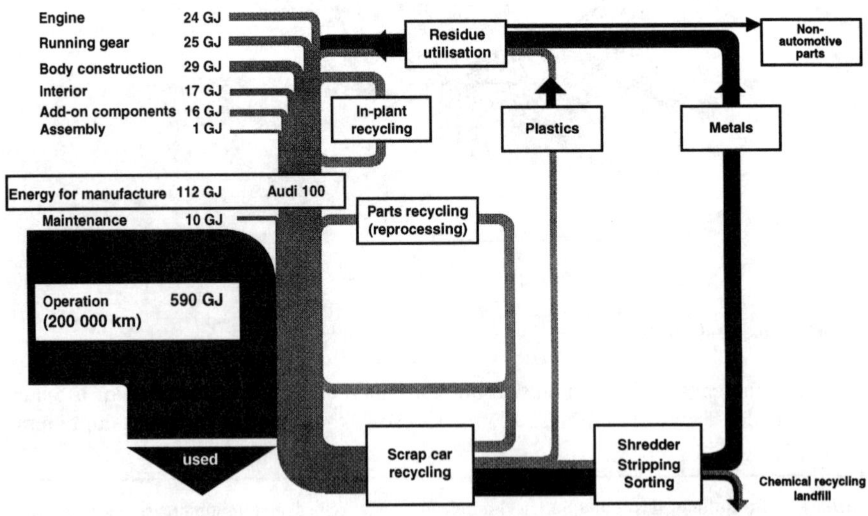

Figure 15 Use of energy for a mid-size car in aluminium intensive construction (secondary aluminium)

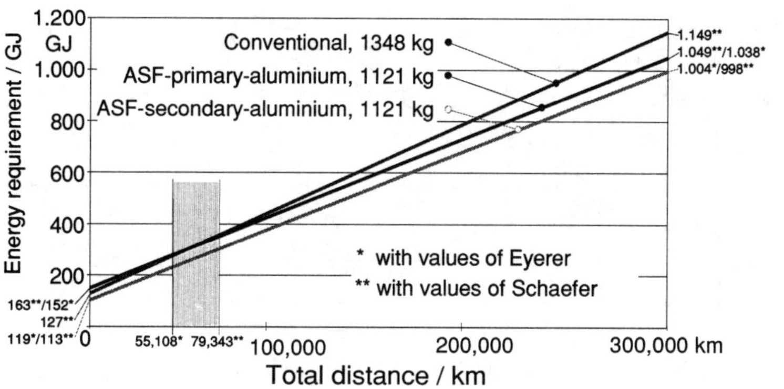

Figure 16 Energy requirement for upper mid-size cars against total mileage

The development of the finished vehicle's energy expenditure for a distance of up to 200,000 km is shown in Figure 16 for several vehicle designs. The car made with new aluminium starts off with a production energy demand of 163 to 152 GJ, the conventional Audi 100 with 127 GJ and the ASF car made with recycled aluminium with 112 GJ. The energy data on production from new material comes from two different independent data bases at the Technical University in Munich and the University of Stuttgart. The differences owing to different processes and geographical locations mean that appropriate margins have to be accorded to the findings. Nevertheless, the following tendencies are clear.

The curves for Audi 100s made of new steel and aluminium intersect at a point between 55,000 km and 79,000 km, ie. the extra expenditure for the high-aluminium content design pays for itself several times over in the course of the vehicle's first service life (eg. four to six times over in a distance of 300,000 km). In the start situation, given a production energy demand in the order of 140 GJ, new aluminium is at a disadvantage of 25 GJ. Over a distance of 300,000 km the overall energy requirement is in the order of 1000 GJ and the original 25 GJ handicap has turned into an advantage of approx. 170 GJ. That means a saving of approx. 17% on an overall energy requirement of approx. 1000 GJ for production and operation. The point of intersection is at 60,000 km. The more recycled aluminium is used the lower the point of intersection is. When the proportion of recycled aluminium used reaches 70% the extra expenditure pays for itself right away, ie. the energy balance sheet figures for the Audi Space Frame vehicle are more favourable than for the conventional design from the very first kilometre on. Similar results apply to CO_2 emission - of importance for its effects on the climate - which must also be considered throughout the course of the entire product cycle (Figure 17). Owing to CO_2 emission in new aluminium production being higher than in the case of steel the point at which the extra expenditure begins to pay for itself is somewhat later, at about 90,000 km, ie. in this case too the extra expenditure pays for itself several times over in the course of a single vehicle service life! The ASF car made with recycled aluminium also starts off in a more favourable position than the conventional vehicle with regard to CO_2 emission.

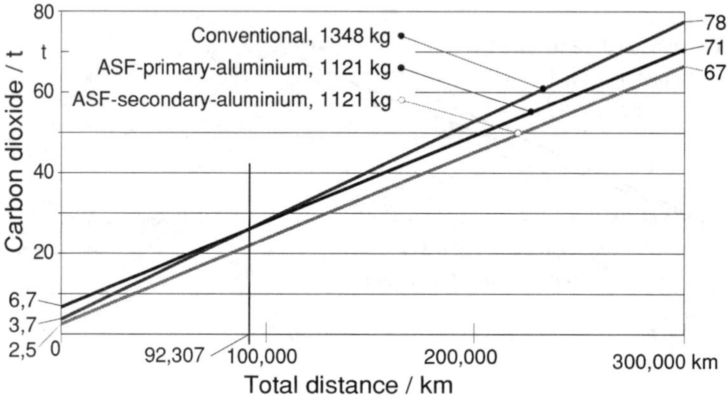

Figure 17 CO_2 emissions for upper mid-size private cars against total mileage

In short, clever lightweight construction compensates several times over for the relatively high energy requirement and the higher CO_2 emission involved in production from new aluminium by comparison with high-steel content vehicles during the first service life of the vehicle and the superior recyclability of aluminium results in a positive balance of energy requirement and

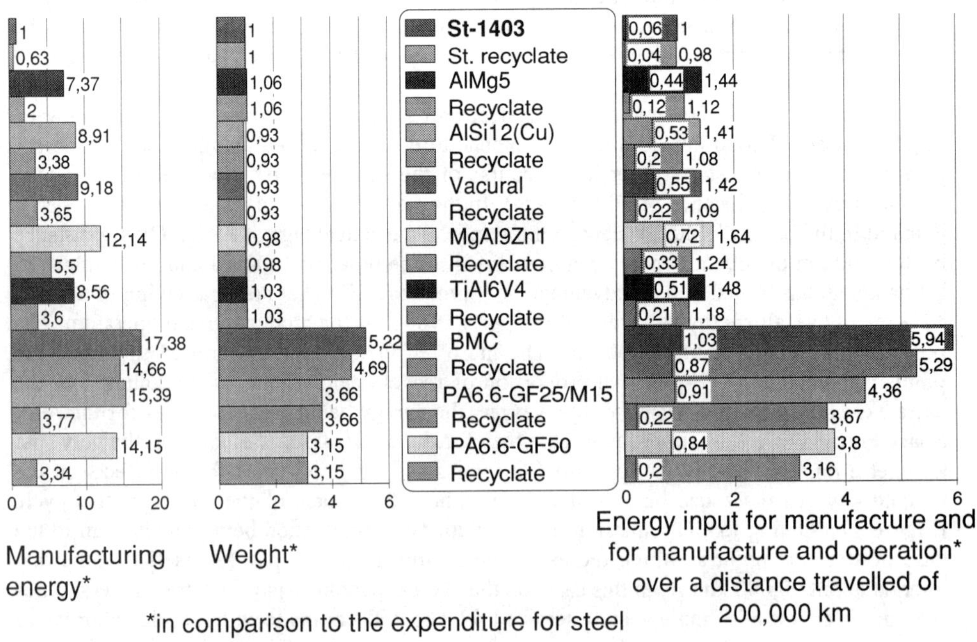

Figure 18 Energy expended with a material selection for tensile/pressure rigidity (V ~ 1/E)

CO_2 emission right at the production stage in the second service life. In a nutshell: aluminium has become an ecologically sound material for vehicle construction!

Classification of lightweight materials by means of general comparison

The expenditure of time and effort involved in drawing up an energy balance sheet in the manner described above with reference to valve bonnets and the Audi Space Frame body is relatively low in comparison to that required for an ecological balance sheet but nonetheless considerable. The things that have to be ascertained and taken into account include the location and extraction of the raw materials, refining them for industrial use, processing them into semis, making these into components and assemblies, use and recycling plus all the transport involved and the general cost accounting allocations for overheads. The need for this multiplicity of details means that an energy balance sheet cannot be drawn up until the production materials have been selected and the design has been worked out in great detail, ie. at a relatively late stage in the development of a product. However, an estimate is often required at an earlier stage. The principle of estimating the suitability of various materials using lightweight construction characteristics combined with particulars of the specific energy demand of typical components made of various materials seems helpful in this context. This method has already been used in a variety of contexts [5 to 13] and will now be discussed with reference to the cumulative energy requirements [1,2,3,4,6] ascertained above for provision of material and production. Figures 18 to 26 are based on the characteristics given in Table 1 which themselves are calculated using the materials values given in Table 2.

Table 1 The characteristics used as a basis for figures 18 - 26

Dimensioning aim	Type of stress	Characteristic	Number
Rigidity	Tension/pressure	$V \sim \frac{1}{E}$	1
	Deflection (w/h = const.)	$V \sim \frac{1}{\sqrt[2]{E}}$	2
Rigidity or stability	Deflection (w = const.) Pressure (w = const.)	$V \sim \frac{1}{\sqrt[3]{E}}$	3
Strength	Tension/pressure	$V \sim \frac{1}{\sigma}$	4
	Deflection (w/h = const.)	$V \sim \frac{1}{\sqrt[3]{\sigma^2}}$	5
	Deflection (w = const.)	$V \sim \frac{1}{\sqrt[2]{\sigma}}$	6
Energy storage	Tension/pressure/ deflection	$V \sim \frac{E}{\sigma^2}$	7
	Width: w Thickness: h	Volume: V Modulus of elasticity: E Strength: σ	

Table 2 The material values used to calculate the characteristics in figures 18 - 26

Boundary conditions						
Reference distance	200.000		km			
Weight factor	0,60		ltr./10^4 km·kg			
Energy used, petrol	36,22		MJ/ltr			
Material properties						
	Manufacturing energy		Density	Modulus	Strength	
	Primary production	Recycling	ρ	E	σ static Rp0,2	σ dynamic (10^7 cycles)
Materials	kJ/cm^3		g/cm^3	GPa	MPa	
St-1403	220	137	7,85	210	220	
R-St-37-2	228	115	7,85	210	220	170 (Rbw)
50CrV4	263	158	7,90	210	1.500	440 (Rzdw)
AlMg5 (import structure)	504		2,64	66,6	110	
AlMg5 (hydroenergy)	435		2,64	66,6	110	
AlMg5 (Western world mix)	543		2,64	66,6	110	
AlMg5 (German mix)		136	2,64	66,6	110	
AlMg5 (Fossil energy)	713		2,64	66,6	110	
AlSi12(Cu)	695	264	2,65	76	170	
"-Vacural	716	284	2,65	76	170	
AlMgSi0,8	396	70	2,64	72		150 (Rbw)
MgAl9Zn1	598	271	1,75	48	130	120 (Rbw)
TiAl6V4	1010	425	4,43	115	830	580 (Rzdw)
BMC (UP-GF30)	178	150	1,95 / 1,755	10	50	
PA6.6-GF25/M15	150	37	1,30	9,5	155	
PA6.6GF50	194	46	1,57	13,3	200	
EP-CF67(HT)	389		1,52	132	1.910	1.337 (Rbw)
EP-CF68(HM)	410		1,60	207	1.100	770 (Rbw)
EPAr-63	340		1,38	75,8	1.380	950 (Rbw)
EP-GF69	103		1,96	43,8	1.100	350 (Rbw)

The seven characteristics indicate the volume of the material concerned needed to achieve the intended dimensions. Taken together with the specific energy demand or the density the characteristic will give the relative energy demand or relative weight or lightweight construction characteristic of a component of the required dimensions. An estimate using the characteristic will only work if assumptions regarding dimensions, stress and boundary conditions (second column of Table 1) are in fact accurate. This, of course, requires a certain amount of intuition and experience. Technical influences such as very low wall thicknesses cannot be taken account of. The method also falls down when geometrically dissimilar alternatives are involved, which will usually be the case with fins, hollow sections or different jointing techniques. Despite all these limitations past experience shows that the method generally produces useful information for the assessment of performance.

As recyclate is regarded as being as good as the new material, the characteristics for new and recycled material in Table 2 are identical. BMC recyclate performs as well as the new material while being 10% lighter (a proportion of the filler is replaced with a mixture of filler and polyester resin), the densities involved being 1.95 and 1.76. Table 2 also contains the essential boundary conditions of these considerations. In addition to the distance travelled in the course of the service life of the vehicle (200,000 km) these are the relationship of fuel consumption to weight 0.6 l/100 km · 100 kg) and the energy value of the petrol (36.22 MJ/l including 4 MJ/l for supply).

Consideration of the valve bonnet materials with regard to performance and energy demand under static stress

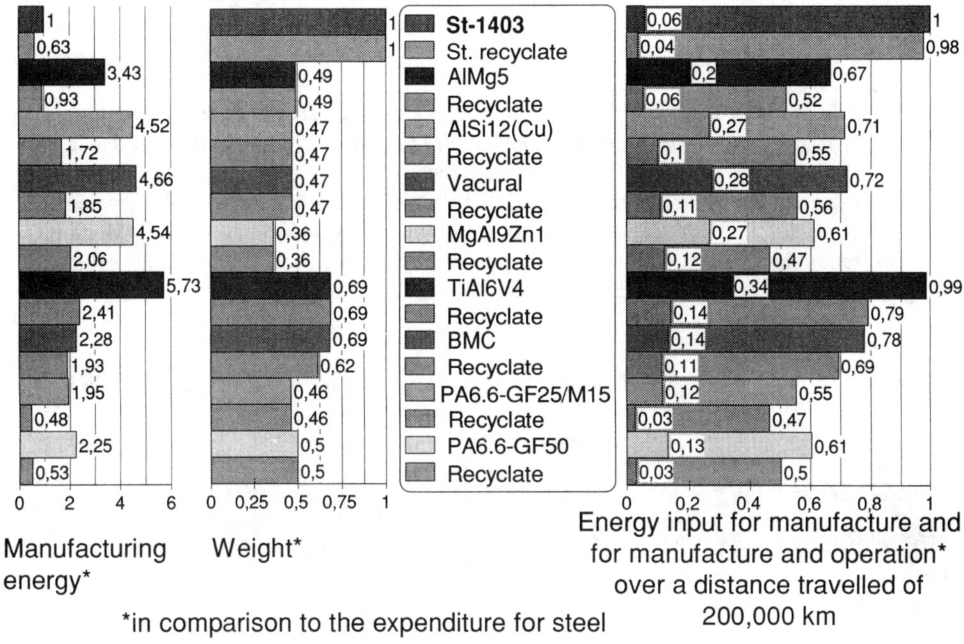

Figure 19 Energy expenditure with a material selection for bend rigidity or stability ($V \sim 1/\sqrt[3]{E}$, w = const.)

In Figures 18 to 20 the characteristics are discussed with reference to valve bonnet materials (and it becomes evident here that the valve bonnet is an example of a component that cannot be dealt with using the principle of characteristics). Figure 18 shows the energy expended with a material selection for tensile/pressure rigidity in comparison to the energy expenditure for steel. The characteristic used as a basis is the first one in Table 1 ($V \sim 1/E$). Following the materials key in the centre, the first column shows cumulative energy expenditure in production. Note: tensile/pressure rigidity produced with MgA19Zn1 requires 12 times the amount of energy that new steel does; recycled steel reduces the requirement to 63% of that for new steel. The second column gives the required weight of the component, once again compared to steel as the reference material. In the case of the lightweight metals there can be scarcely any savings on weight, given the stresses involved. The energy expenditure for production and the overall energy expenditure for production and operation for 200,000 km are given on the right. The reference value here is the overall expenditure for steel. For instance, in the case of steel energy expenditure in production amounts to 6% of the overall expenditure. Production with recycled steel requires 4% of the overall requirement for new steel. As new and recycled steel both weigh the same, the fuel demand is also the same and the 2% difference in production requirement is carried through to the overall balance sheet. Production with BMC requires 17 times the energy expenditure needed with steel: the production requirement alone is as high as the overall requirement for production and operation with steel. The overall BMC requirement would be 6 times that of steel. In short, none of the valve bonnet materials is better suited for tensile/pressure rigidity than steel: it has the lowest production energy requirement and the differences in weight offered by the light metals are insignificant.

Analagous to Figure 18, Figure 19 shows the energy expended on attaining a set bend rigidity or stability in comparison to the energy expenditure for steel. Volume is in accordance with the characteristic 3 (V ~ $1/^3\sqrt{E}$). Of the new materials steel is once again the material with the lowest production energy requirement. Recycled aluminium is better than steel, recycled steel is even better and the best alternative is provided by the recycled thermoplastics which have half the requirement of new steel. All the alternatives to steel offer substantial savings in weight of between one and two thirds when looked at from the point of view of bend rigidity.

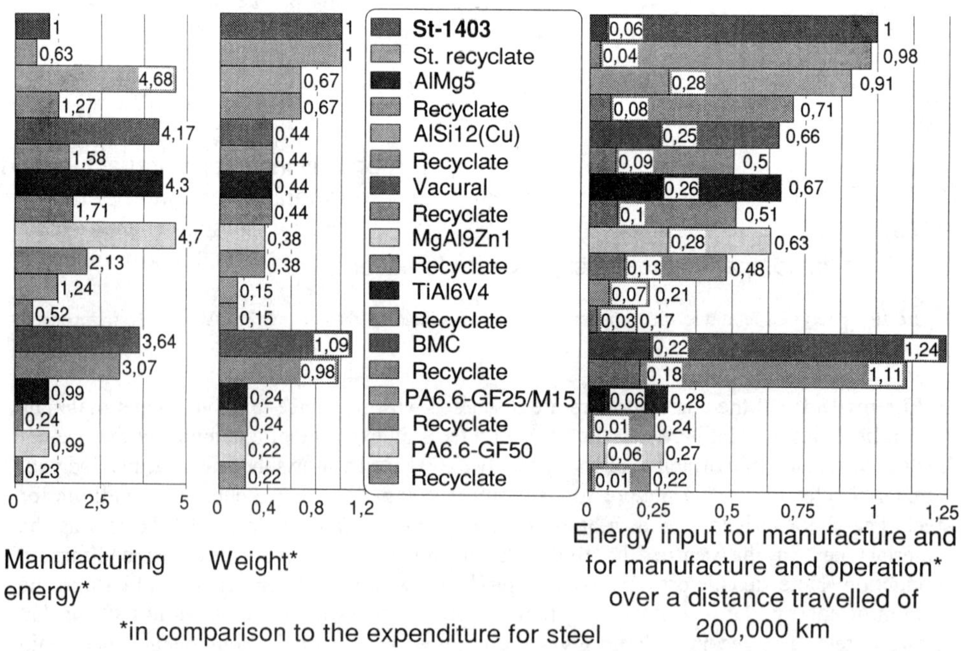

Figure 20 Energy expenditure with a material selection for tensile/pressure strength (V ~ 1/σ)

Its high production requirement makes the overall balance for new titanium no more favourable than for steel. Magnesium, the reinforced 14" thermoplastics and the aluminium versions make a very advantageous showing. The 40% recycling bonus in production with recycled steel has, at 2%, practically no effect on the overall energy requirement, as production of the heaviest version only accounts for a twentieth of its overall requirement.

Figure 20 shows the energy expenditure that would be necessary using the same materials if strength corresponding to characteristic 4 (V ~ 1/σ) had been the aim. The advantage of steel is in its production requirement; however, the weight savings offered by the other materials with the exception of BMC are so considerable that they make a far better showing in the overall balance sheet.

In view of the advantages offered by the non-ferrous materials with regard to bend rigidity/stability and tensile/pressure strength the matter of whether the wall thicknesses indicated by the characteristic values are in fact technically practicable ought to be looked into on the basis of actual components. In the case of the valve bonnets there were certain limitations. Figures 19 and 20 provide impressive evidence of how wrong it can be to put the production energy requirement in the forefront of deliberations concerning environmental suitability.

Consideration of dynamically efficient materials with regard to performance and energy requirements

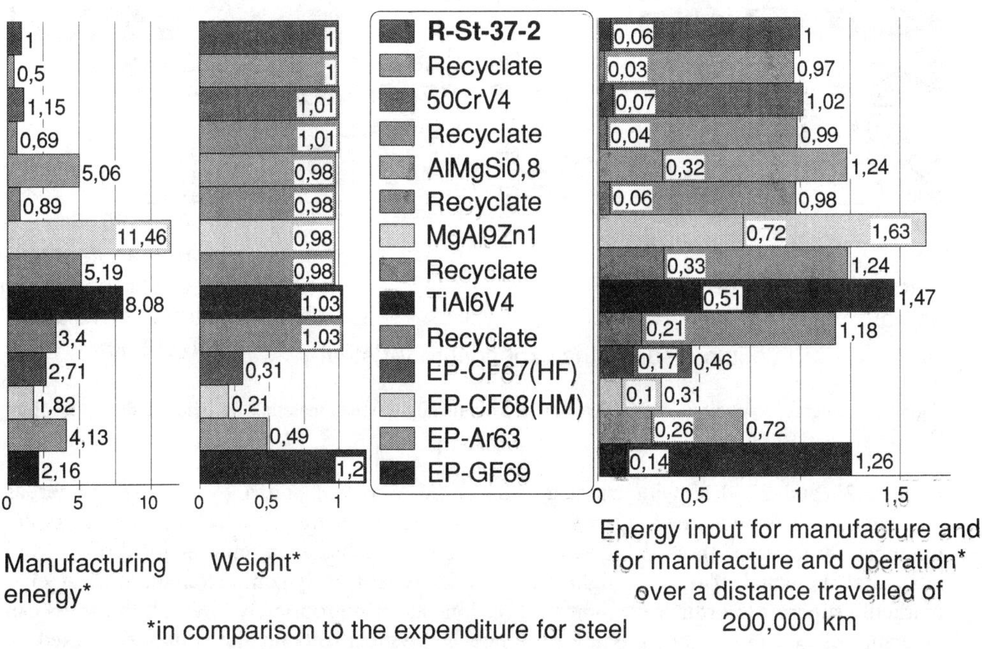

Figure 21 Energy expenditure with a selection of dynamically efficient materials for tensile/pressure rigidity (V ~ 1/E)

Building components to resist dynamic stress is more often an issue in vehicle construction than resistance to static stress. Figures 21 to 26 are based on the characteristics given in Table 2. In contrast to the discussion on materials for valve bonnets, materials have been selected here which are more suitable to withstand dynamic stress and for which figures on dynamic strength are available. The primary reference material is steel. In addition to carbon steel of the kind used for girders the quality 50CrV4 is taken as an example of a high-alloy steel. Figure 21 shows the energy expenditure necessary to achieve tensile/pressure rigidity in accordance with characteristic 1 (V ~ 1/E) for these materials. Once again, steel is outstandingly well suited to stress of this kind. To achieve energy savings one has to turn to high-performance carbon or aramide fibres, taking into consideration the fact that the low resistance to pressure of aramide fibres is not dealt with in this account.

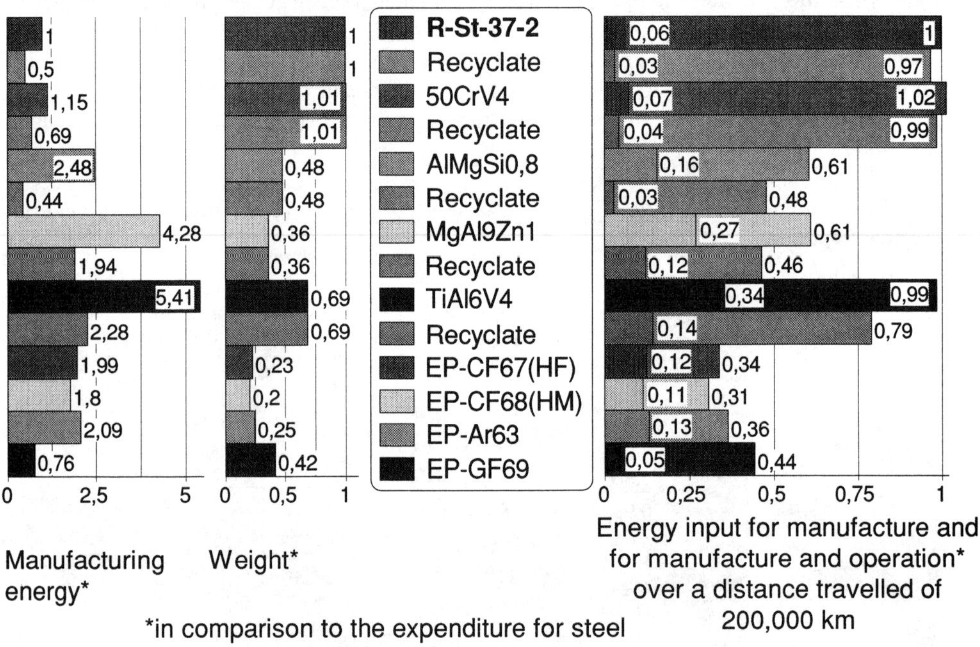

Figure 22: Energy expenditure with a selection of dynamically efficient materials for bend rigidity or stability ($V \sim 1/^3\sqrt{E}$, w = const.)

Figures 22 and 23 deal with characteristics 3 ($V \sim 1/^3\sqrt{E}$) and 4 ($V \sim 1/\sigma$). As far as characteristic 3 is concerned, apart from titanium the alternatives to steel are more suitable: the most attractive of these is high-modulus carbon fibre but even unidirectional glass fibre is at least the equal of the best light metals. With regard to dynamic strength alloy steel is markedly superior to ordinary carbon steel (as long as an appropriately low wall thickness can be achieved). Of the metals, the titanium alloy is the most suitable and is only surpassed by the fibre reinforced plastics - of which it must also be said that even unidirectional glass fibre performs better.

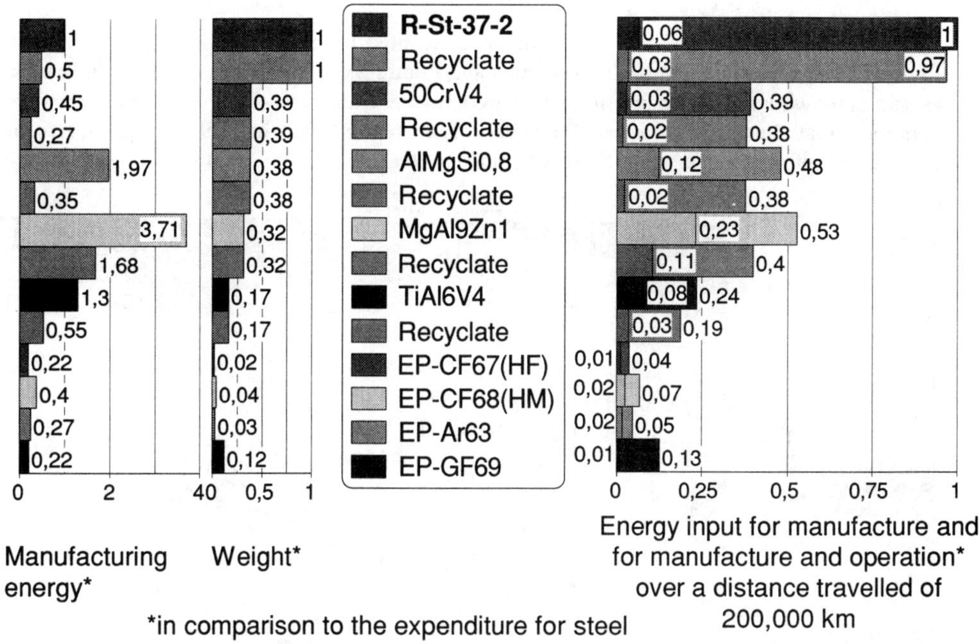

Figure 23: Energy expenditure with a selection of dynamically efficient materials for tensile/pressure strength (V ~ 1/σ)

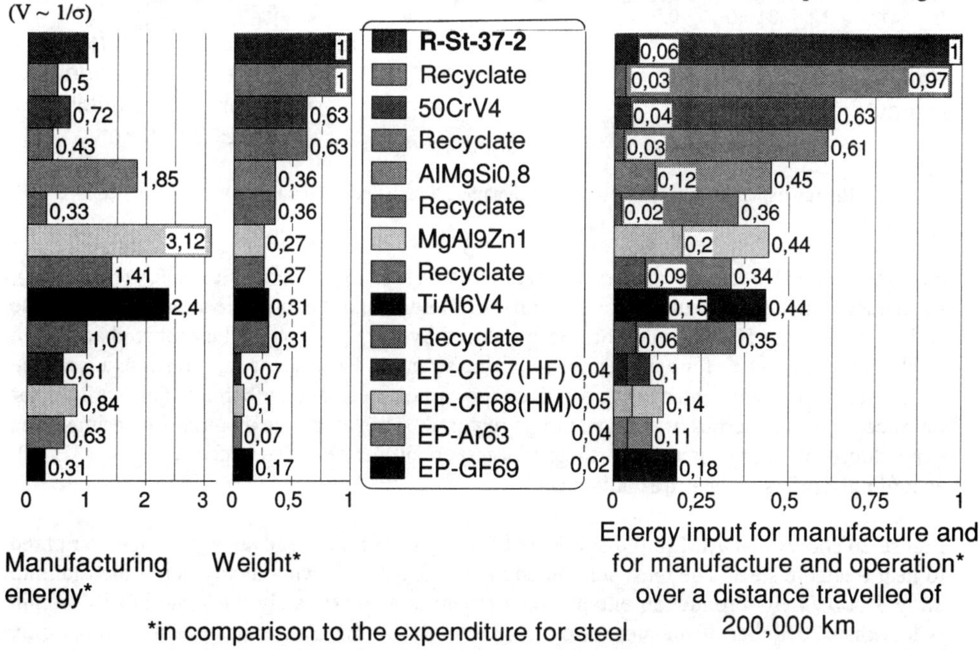

Figure 24 Energy expenditure of a selection of dynamically efficient materials for bend rigidity ($V \sim 1/^{2}\sqrt{\sigma}$, w = const.)

Reaction to bending stress at a set width is represented by characteristic 6 ($V \sim 1/^2\sqrt{\sigma}$) in Figure 24. The main difference between this and Figure 23 is the advantage the light metals, as materials with a low density, have when compared with alloy steel. Given the boundary condition that the cross-sections of the various materials must be similar (eg. square or circular or with the same width/thickness ratio) so that

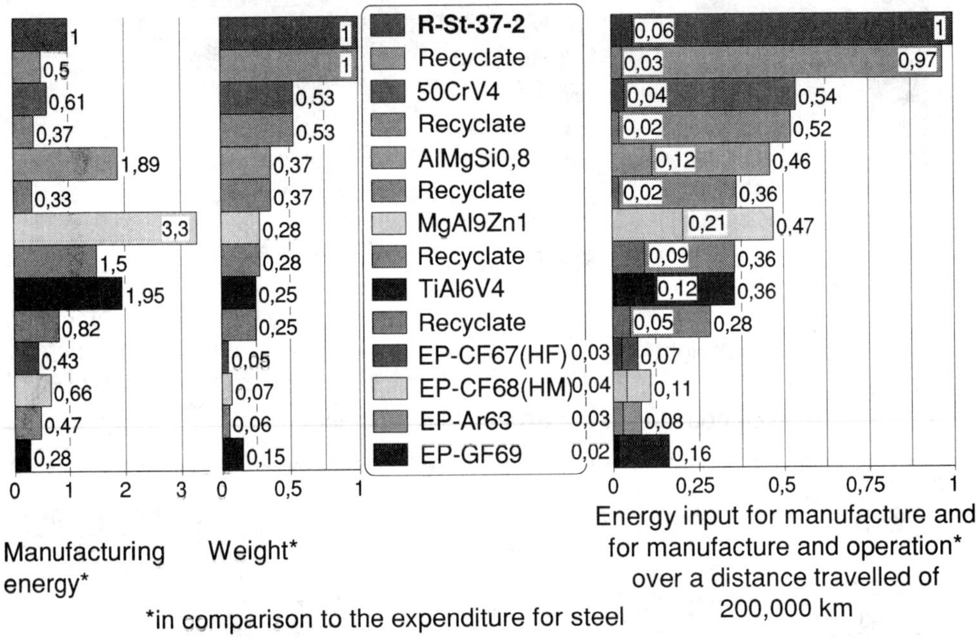

Figure 25 Energy expenditure with a selection of dynamically efficient materials for bend rigidity ($V \sim 1/^3\sqrt{\sigma^2}$, w = const.)

characteristic 5 ($V \sim 1/^3\sqrt{\sigma^2}$) applies, the most suitable solution is the titanium alloy (Figure 25). In characteristics 5 and 6 the factor of approx. 2.5 indicating the dynamic strength ratio of the carbon steel to the 50CrV4 alloy has been reduced by the square and cube root so that carbon steel remains attractive for applications requiring strength. However, in applications where the difference in strength comes to the forefront the carbon steel lags far behind. This applies, for instance, to the properties of energy storage required, for example, in material for springs. The expenditure of energy necessary to produce such properties, characteristic 7 ($V \sim E/\sigma^2$), accordingly refers to heat treatable steel.

Figure 26 shows that AlMgSi0,8 and MgAl9Zn1 have certain disadvantages when compared to heat treatable steel. The most suitable metal in the series is without any doubt the titanium alloy TiAl6V4. Its production energy requirement is approximately the same but its weight is less than 20% that of the steel alloy. Accordingly, in the overall balance it requires only 25% - or, in the case of recycled titanium, 20% - of the energy expenditure necessary for steel. These are lines it would be worthwhile to develop along, particularly as the formerly very high price of titanium has fallen drastically in the last few years. The figures used for production energy requirements (1010 kJ/cm^3 for new and 425 kJ/cm^3 for recycled titanium) were worked out from published data with a relatively wide spread by *Huber*[4]. They represent an average of a fairly wide spread. In actual practice the figure would probably be below this average, as the figures for energy requirements were higher at the time of publication. The conclusion may be drawn that the processes employed in industry have improved. Research ought to pay off in this connection. This also applies to fibre composite materials whose potential is substantially greater than even titanium alloy. What is needed here are strategies of design and technology that will permit at least a part of this potential to be exploited in engineering practice.

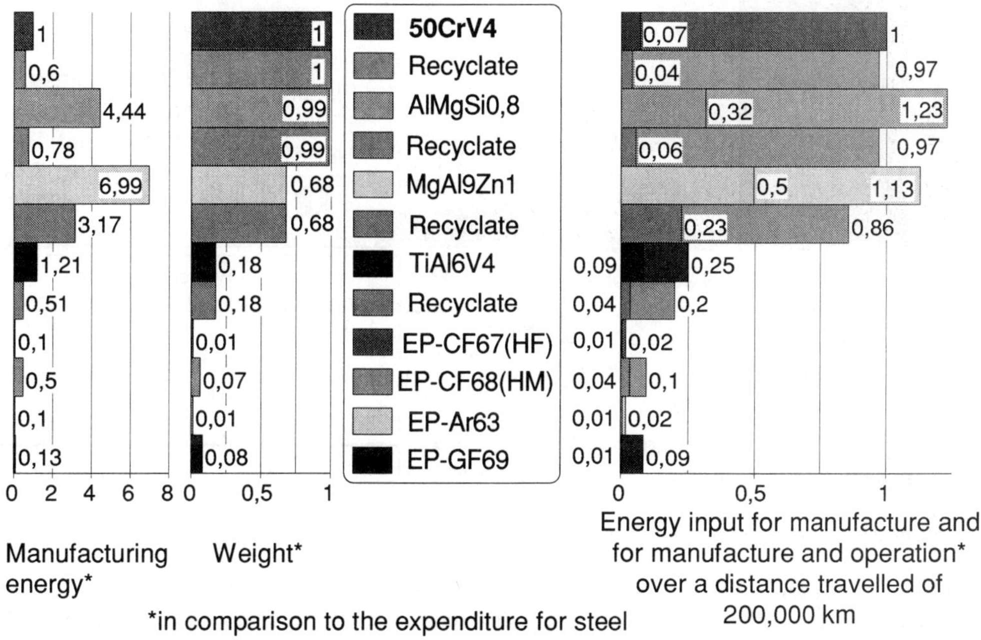

Figure 26 Energy expenditure with a selection of dynamically efficient materials for energy storage ($V \sim E/\sigma^2$)

Alternatives to recycling

Both ecological and energy balance sheets point to better alternatives with regard to materials, strategies, processes and locations [14, 15]. A comparison of the alternatives to recycling plastics from scrap cars would appear to be of interest in this connection. The background is as follows. The car industry, ie. the manufacturers and suppliers of automobiles together with their materials suppliers support the campaign to reduce quantities of waste, which has now also been joined by politicians, in accordance with the sensible order of priorities "Avoid the creation of waste. If it can't be avoided, recycle it. Disposal only as a last resort". A common effort by the initiators of the campaign has led to materials being recycled that never were before.

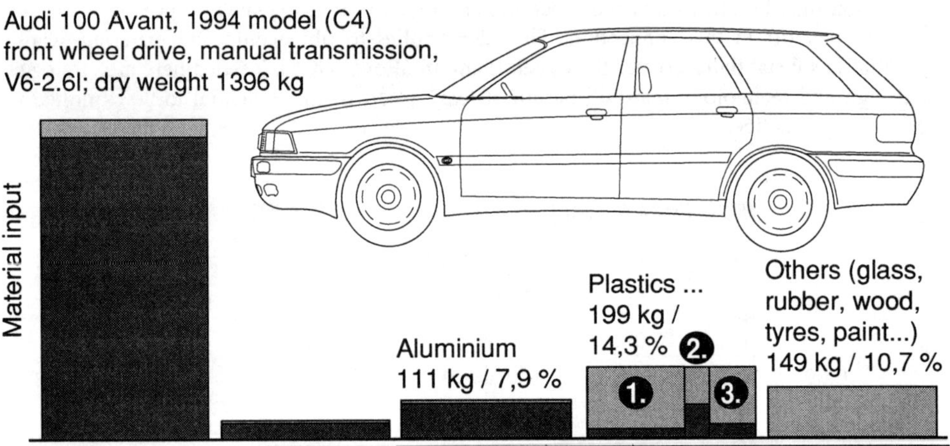

Plastics .../kg	Total utilisation	Recyclate shares Production	scrap car	Total Recycl.
1. Plastic components	116,5	11,4	3,7	15,1kg/13%
2. Textiles,molded fibre parts...	27,15	1	11,7	12,7kg/46,7%
3. Damping mat, underseal, seam seals	55,9	11	0	11kg/19,7%
	199,6	23,4	15,4	38,8kg/19,4%

Figure 27 Material input and use 1993, using the example of the Audi 100 Avant

Taking the example of the Audi 100 Avant, Figure 27 shows that the release of organic materials (recyclates of production waste and scrap cars) for recycling has already reached considerable proportions. On average and leaving tyres and paint out of account, 19.4% of organic materials have been released for recycling. In the case of plastics alone the proportion is 13%, of which 3.2% is itself recycled old components. This proportion is high and will continue to grow. The question of how high the proportion of material recycled should reasonably rise is still under discussion by the experts. At a work shop[7] the figure of 10% - 25% was discussed and the conclusion reached that it was still too early to talk about definite figures. There was, however, agreement on the point that the target of recycling 50% of plastic from scrap cars set by the federal government was unrealistically high. Even if the highest figures entertained by the experts, ie. 25% - 30%, were attainable and desirable the alternatives to materials recycling could not be dispensed with, as is illustrated in Figure 28.

[7]RE WORK SHOP 2; Bonn, 12.11.93, organized by Maack Business Services.

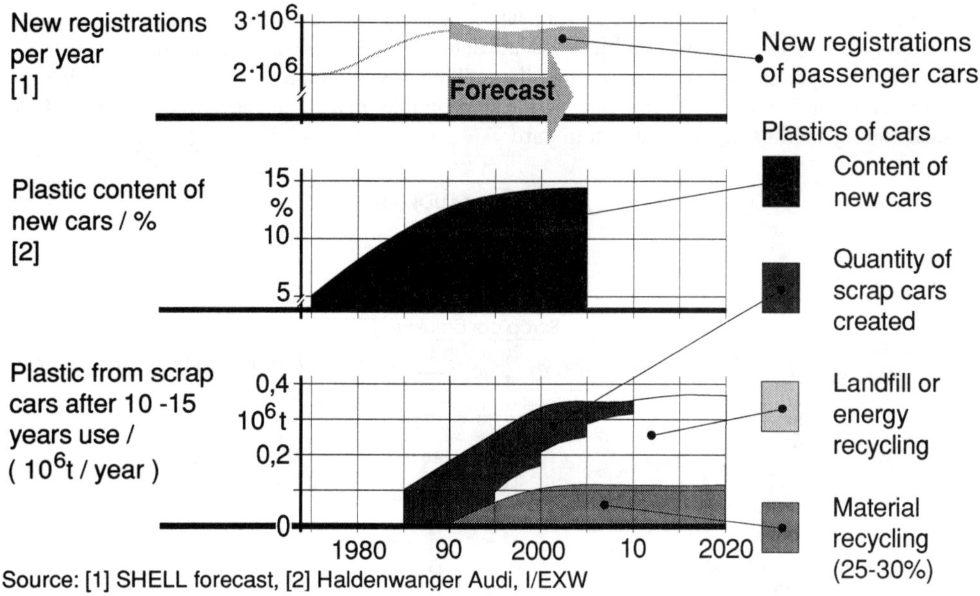

Figure 28 Incidence and reuse of scrap car plastics in the Federal Republic of Germany, 1989 (without the former East Germany)

In view of the quantities of plastic anticipated from scrap cars the amounts in need of disposal would continue to grow even with a recycling rate of 25% - 30%. In other words, materials recycling will not solve the waste disposal problem by itself and, despite all the successes in this field, the alternatives are in fact more important! Materials recycling must be supplemented by chemical recycling and conversion to thermal energy. The energy balance sheet shows the good sense of efforts to utilize scrap car plastics either thermally or by way of materials recycling rather than disposing of them in landfills.

Würstle/Schaefer[8] dealt with this topic, taking the example of an Audi 80 (IFE Technical University in Munich). *Woyke/Schaefer*[1] had previously calculated a cumulative energy expenditure of 110 GJ (including indirect energy consumption) for the production of this car. The car weighs 1217 kg. The scrap dealer removes 287 kg for re-use, of which 6.5 kg are plastics (4% of the quantity of plastics originally used in production). 930 kg are sent to the shredder. There and in the following sorting and separation lines a total of 239 kg of waste results which has hitherto been disposed of in a landfill (Figure 29). The study discusses landfill storage (column S1 in Table 3) and three alternative methods of treatment (S2 to S4). In the S2 alternative the calorific value of the waste is exploited. S3 involves recycling a

[8]Würstle, J.: Der kumulierte Energie-Aufwand zur Entsorgung eines Mittelklasse-PKWs am Beispiel des Audi 80 (The cumulative energy expenditure involved in the disposal of a mid-size car, taking the Audi 80 as an example) graduate diploma thesis at Lehrstuhl für Energiewirtschaft und Kraftwerkstechnik, Technische Universität München 1993

proportion viewed by the experts as very large[7], namely 40 kg out of 147 kg, and disposing of the rest in a landfill. S4 combines the advantages of S2 and S3, dismantling the cars to facilitate recycling and exploiting the thermal energy contained in the fraction that is unsuitable for recycling. In the case of the two options involving recycling (S3 and S4) a dismantling facility replaces the scrap yard.

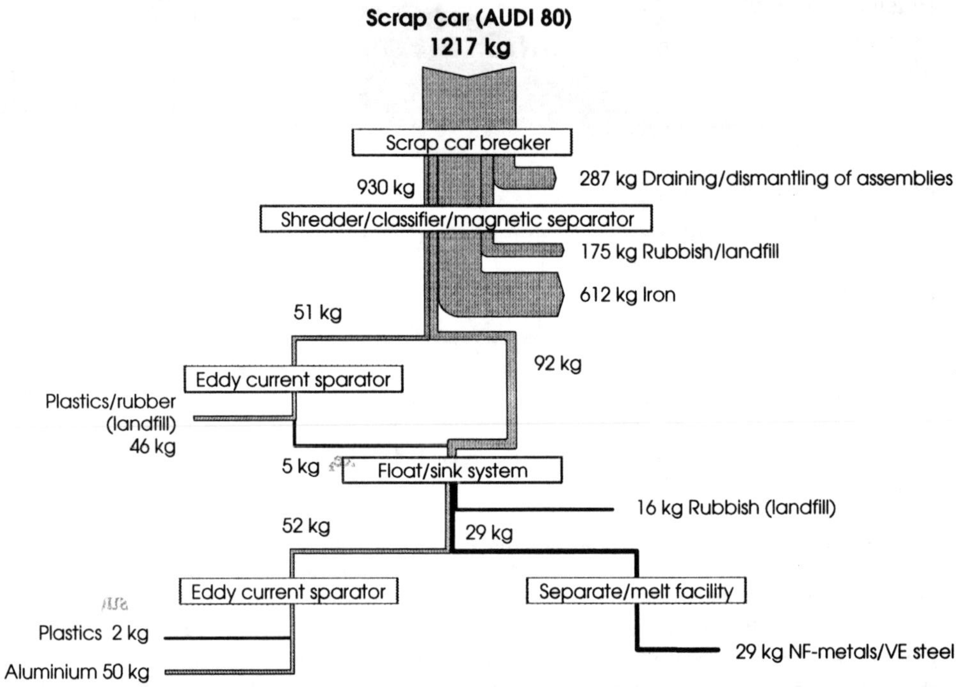

Figure 29 Material flow for a passenger car of type Audi 80 in the current stages of scrap car disposal (Source: IFE, Würstle/Schaefer, 1993)

Table 3 provides the following data. Disposal of scrap cars (see also Figure 30) has hitherto required an energy expenditure of 0.7 GJ per car (according to other sources 2 GJ per car[2]). 0.7 GJ increases the cumulative energy expenditure per vehicle from 110 GJ to 110.7 GJ, ie. by 0.6%. By producing 2.14 GJ of energy thermal exploitation of the residual waste would reduce the cumulative energy expenditure per vehicle from 110 GJ to 107.86 GJ, ie. by 2%. If instead the above-mentioned quantity of plastics, ie. 27% of that used, were recycled the energy bonus would be higher, namely 2.3 GJ. This would reduce the cumulative energy expenditure per vehicle to 107.7 GJ, ie. by 2.1% rather than 2%. The combination of both methods, recycling and thermal exploitation, would produce an energy bonus of 4.25 GJ, ie. a 3.9% reduction of the cumulative energy expenditure per vehicle to 105.75 GJ.

Looking at current practice, the differences are as follows. Thermal exploitation alone would bring a gain of 2.8 GJ per mid-size car. Maximum recycling would bring a gain of 3 GJ, ie.

not quite 10% advantage over the 2.8 GJ gained by thermal treatment. The combination of thermal exploitation and recycling would bring a gain of 4.95 GJ. That would be a 75% advantage over thermal treatment alone and 65% over recycling alone.

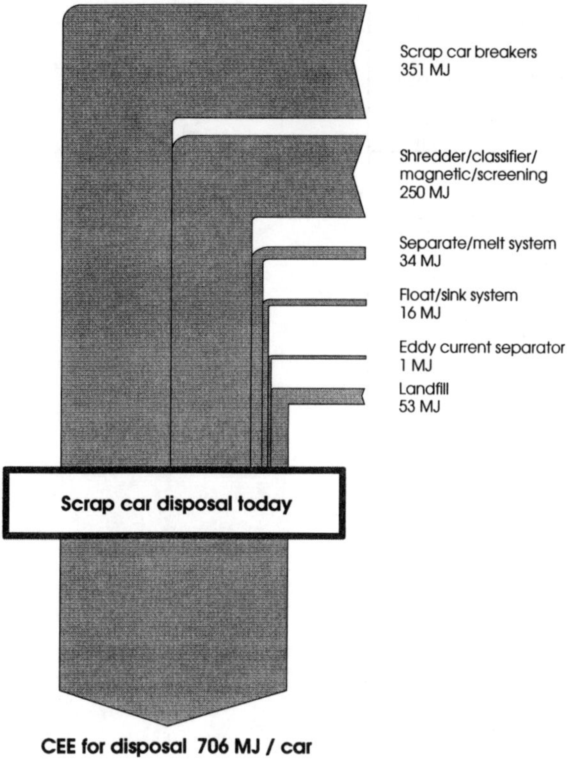

Figure 30: Cumulative energy expenditure for disposal of a passenger car of type Audi 80 in the current stages of scrap car disposal (Source: IFE Würstle/Schaefer, 1993)

While recognizing that politicians are taking the right path in giving recycling priority over thermal conversion it cannot be ignored that the advantages are smaller than many had expected. It would be completely wrong to dispense with the alternatives on the basis of the possibilities offered by recycling. Both ways must be pursued and the alternatives to recycling are in fact of greater importance. Despite the successes of the engineers the 50% recycling target set by the politicians is unrealistically high and would, even if achieved, be insufficient to solve our waste disposal problems.

Table 3: Energy balance for various alternatives of residue utilisation (data in GJ/car)

		Alternatives for residue utilisation or disposal			
		S1	S2	S3	S4
		Current disposal practise	**Fluidized bed incine-ration (Thyssen)**	**Dismantling with com-ponent re-cycling**	**Dismantling with com-ponent re-cycling and subsequent fluidized bed incine-ration**
		Energy data in GJ/car			
Z1	Scrap car breaker	0,35	0,35		
Z2	Dismantling			0,73	0,73
Z3	Shredder	0,30	0,30	0,28	0,28
Z4	Thermal utilisation facility		0,75		0,52
Z5	Deposition	0,05	0,03	0,04	0,02
Z6	**Total energy for disposal** (total of Z1 to Z5)	**0,70**	**1,43**	**1,04**	**1,54**
Z7	Credit for energy recovery or saving by recycling		3,57	3,34	5,78
Z8	Net energy cost for dis-posal path (Z6-Z7)	0,70	-2,14	-2,30	-4,25
Z9	**Advantage over current disposal method** (Z6S1-Z8)	**0**	**2,84**	**3,00**	**4,95**

Summary and conclusions

Considerations of useful life cycle can - sometimes suprisingly - lead to different and ecologically more meaningful results than at first anticipated. Both the ecological and the energy balance sheet often lead to very wide ranging results. Much less expenditure of time and effort is involved in the setting up of an energy balance sheet than in the drafting of complete ecological balance sheets. Our estimates and comparisons so far, using examples taken from the automobile industry, have led us to believe that the results obtained from both kinds of balance sheet agree to a large extent. A critical eye should, however, be kept on the limitations of the usefulness of such results. In the case of products whose use accounts for less of their overall energy requirement than automobiles (eg. stationary machinery) the answers might well be entirely different.

In any event the drawing up of overall balance sheets covering the entire cycle of a product should become standard practice. That also applies to the alternative materials and processes discussed in this paper. Informative statements on other topics, such as the following, are also desirable.

- The use of PVC as compared to its alternatives.
- The use of tropical hardwoods specially grown on plantations rather than harvested at random in the rain forests.
- Regenerable raw materials and energy resources rather than fossil fuels.
- A world-wide comparison of production locations.
- The effects and ramifications of different laws and regulations world-wide. Notions of ethics and morality.
- The effects of the mobility of the individual compared to other scenarios.

Owing to these disparate and sometimes poorly defined criteria throughout the world, the energy balance sheet will always provide more precise aids to practical decision making than the ecological balance sheet. In the case of complex structures, too, such as those involved in automotive engineering it is to be expected that the energy balance sheet will serve as a more effective decision making tool.

The ecological balance sheet will nonetheless remain indispensible wherever scarcity of resources or difficult process techniques are envisaged from the start - for example where dioxin/furan/radioactivity are involved or decisions with far-reaching consequences have to be made, eg. the choice between nuclear or fossil fuel power stations.

References

1 N.N.: Ökobilanz - Definition des Umweltbundesamtes; AVK-Aktuell - Periodikum der Arbeitsgemeinschaft Verstärkte Kunststoffe 10/93 (cological balance sheet - definition by the Federal Department of the Environment) AVK-Aktuell - periodical of the Reinforced Plastics Working Party 10/93

2 Böhm, E. and Walz, R.: Ökobilanzen: Was kann die Fraunhofer-Gesellschaft tun? Fraunhofer Jahresbericht ´92, p. 83 (Ecological balance sheets - What can the Fraunhofer-Gesellschaft do? Fraunhofer Annual Report ´92)

3 Helmüller, R.: Ökobilanzen geben keine einfachen Antworten; VDI-Nachrichten Nr. 42, 22.10.93, p. 28 (Ecological balance sheets give no simple answers)

4 Reiter, K.: Praxisverbrauch verschiedener Leistung/Gewicht-Klassen in Abhängigkeit des Fahrzeuggewichts, Verbrauchsdaten basierend auf ADAC- Angaben in ADAC-Motorwelt 12/1990 (Vehicle weight related fuel consumption in actual practice of different performance/weight classes. Data based on information published by ADAC)

5 Haldenwanger, H. G., Schäper, S.: Die "Energiekette" im Lebenszyklus eines PKW; (The "energy chain" in the life cycle of an automobile), paper given at the 11th VDI-VW joint conference "Neue Konzepte für die Autoverwertung", Wolfsburg, VDI Berichte 934, ISBN 3-18-090934-X

6 Mast, P., Schäper, S, and Wagner, D.: Conservation of resources by recycling plastic parts from automobiles. Paper given at the Annual Conference Plastics in Automotive Engineering, Mannheim 1991, VDI-Verlag, ISBN 3-18-401144-5, p. 25/64
7 Haldenwanger, H.-G. and Schäper, S.: Ressourcenschonung durch Recycling; ATZ93 (1991) 11, p. 676/686 (Conservation of resources by recycling)
8 Haldenwanger, H..-G., Schäper, S.: Light cars: a rational approach to the best use of resources. Paper given at the seminar ATA-MAT 90 at the XIII Fisita Congress, 7 - 11 May 1990, Turin.
9 Haldenwanger, H.-G., and Schäper, S.: Ressourcenschonung durch Leichtbauwerkstoffe? Beitrag zum Kongreß "Technik und Politik - Kein Ausweg aus der Fortschrittsfalle?" der Bayerischen Landtagsfraktion "Die Grünen", 12./14.7.90, (München Conservation of resources with lightweight materials? Paper given at the congress "Technology and politics - no way out of the progress trap?" organized by the Bavarian state parliamentary "Green" party in Munich on 12/14.7.90)
10 Haldenwanger, H.-G., and Schäper, S.: Neue Werkstoffe und Verbundstrukturen im Automobilbau; (New materials and composite structures in the car industry). Paper given at VDI Werkstofftag, München, VDI-Verlag ISBN 3-18-090734-7, p. 1/37
11 Haldenwanger, H.-G. and Schäper, S.: Faserverbundwerkstoffe aus der Sicht des PKW-Entwicklers; Plastverarbeiter 34 (1983) 8, p. 677/681 (Fibre composite materials from the point of view of automobile designers)
12 Schäper, S.: Kunststoffeinsatz im Automobil hilft Energie sparen; Kunststoffe 70 (1980) 6, p. 319-324 (Using plastics in cars helps save energy)
13 Schäper, S.: Energieeinsparung durch Verwenden von Kunststoffen im Automobil; (Saving energy by using plastics in cars). Paper given at the 27th SKZ plastics conference "Energy saving in plastics processing and applications" in Würzburg on 6.12.1979
14 Eyerer, P., Dekorsy, Th., Schuckert, M. and Hesselbach, J.: Produkte, Werkstoffe, Umwelt -Ganzheitliche Bilanzierung von Bauteilen, Ingenieurwerkstoffe 3 (1991) Nr. 7/8, p. 52-57 (Products, materials, environment - drawing up a holistic balance sheet of components)
15 Eyerer, P., Dekorsy, Th., Schuckert, M. and Pfleiderer, I.: Ganzheitliche Bilanzierung von Bauteilen als Entscheidungshilfe für die Auswahl von Produktentwicklungen und Recyclingverfahren; (Holistic balance sheet of components as an aid to decision making when selecting product development plans and recycling processes), Paper given at the 11th VDI-VW joint conference "Neue Konzepte für die Autoverwertung", VDI Bericht 934, ISBN 3-18-090934-X

The use of natural fibres in vehicle components

Wolfgang R. Wittig

Until around seventy years ago, natural materials were practically the only alternative for the production of all types of objects and technical products. Textiles, rope, canvas and paper were manufactured from domestic natural fibres such as flax, hemp, etc., and are, indeed, in some cases still used today. The first plastics were obtained not from mineral oil, but were based instead on milk (casein) or modified cellulose in the form of cellophane or products made of cellulose esters and similar materials. The rise to predominance of synthetic organic materials, and of fibres, in particular, which make possible our high-quality and high-quantity present-day standard of living, started in the 1920s.

The current collapse in prices for engineering and standard plastics (they are currently at 1960s levels) and performance potentials which have been continuously improved in the course of time have in recent years practically completely excluded the use of regenerable raw materials. The increasingly critical discussion in the recent past of subjects such as the greenhouse effect, recycling, and the finite nature of our crude oil reserves has resulted in certain sections of the population, particularly in the industrialized nations, adopting a "back to natural materials" attitude with particular emphasis attaching to regenerable raw materials. A further impetus for the use of natural materials is seen in countries such as Brazil in the form of a great super-abundance of natural fibres and a perennial shortage of foreign currency to use for the purchase of imported oil. These factors were also the points of origin of the first components for commercial vehicles suitable for large-scale production, which will be discussed in more detail later. Social circumstances in these countries have also tended to accelerate the technical exploitation of natural fibres. There is, for instance, at present a mid-term project operated jointly by Daimler-Benz Research Ulm, and UNICEF in the Amazon region aimed at increasing the permanence of the region's population by means of a prolonged program of planting of suitable vegetation and its technical marketing, and thus of ultimately reducing fire-based rain-forest clearance, with all its negative implications.

According to information obtained from the Farmers' Association, however, German agriculture is now also perfectly capable of achieving a positive contribution in the case of flax and, given corresponding fiscal legislation, of rape seed, too. Such a background makes the pursuance of vegetable-based materials a rational proposition. The following is intended to outline the present-day status for the use of (reinforcing) natural fibres in motor-vehicles and to estimate the development potential.

Natural materials

The natural materials of the greatest importance and therefore interest for engineering applications must be subdivided into natural matrices, natural oils and natural fibres.

Natural matrices

Among the natural matrices, which we will not be able to deal with in more detail here, lignin (produced in paper manufacturing) should be mentioned as a substitute resin (for phenolic resin, for instance) or resin additive. In addition, protein (as occurs, for example, in soya-oil production) and starch (as obtained from potatoes or maize, for instance). Also familiar is a PA11 polyamide and a polyurethane (PUR) based on castor oil, as are the cellulose esters (CA, CP and CAB). There are now starch-modified polymers with partial biological degradability available, and completely biologically degradable plastics, such as polyhydroxybutyrate (PHB).

Due to their inadequate property profiles, these polymers are not yet found in automotive applications. In an ideal case, it would be possible to develop a "natural composite" consisting of natural fibres bound by a natural matrix, a material which would satisfy technical requirements and would, at the same time, possess the most controllable possible biological degradability.

And in this context, we should state from the start: Development work in this direction is under way, but a satisfactory result is not yet in sight.

natural fibres

plant fibres:

- vegetable hairs: cotton
- bast fibres: flax, hemp, jute, rami
- hard fibres: sisal, coconut fibre, banana fibre

natural oils

- fuels
(based on ramp seed oil, palm oil or alcohol)
- lubricating oils
(based on ramp oil)
- hydraulic oils
(based e.g. on ramp oil)

natural polymer binders

1) polyamid PA11: based on castor oil
2) polyurethane PUR: polyol based on castor oil
3) cellulose esters CA, CP, CAB:
   - e.g. from wood residues
4) natural fibres pressed with natural binders:
   - lignin, tannin, protein or starch
5) starch modified polymers:
   - starch in addition to PE, PS, PP
   (limited biological degradation)
6) biologically degradable polymers:
   - e.g. PHB, PCL, cellulose-di-acetat starch

Figure 1. Classification of natural materials

Natural oils

Natural oils, such as the castor oil already mentioned, palm oil, or our own domestic rape oil, could, after even only partial completion of certain development projects, already be used in lubricants, fuels and hydraulic fluids, primarily in commercial vehicles. For cost reasons, however, practically all commercial-scale tests have been stopped. The natural oils group is not dealt with in any more detail in this article.

Natural fibres

Natural fibres may be subdivided into vegetable fibres (cellulose), animal fibres (protein) and mineral fibres (e.g. asbestos).The natural fibres are the principal subject of this essay.

The bast and hard fibres used as reinforcement materials possess the greatest potential for use in engineering. Since there is probably not particularly great knowledge amongst plastics specialists of obtainment of raw materials from tropical and European natural fibres, we will briefly outline the most important stages, using an example. Bast fibres (also referred to as soft fibres) are found in the stems of herbaceous plants such as flax, for instance. Hard fibres are obtained from leaves, such as those of the sisal and banana plants. Some vegetable fibres, such as cotton and flax, are practically pure cellulose fibres. Others, such as jute, have a higher lignin content. This lignin is located between the cellulose structures, like an adhesive. The cellulose is responsible in particular for strength, and lignin mainly for rigidity, and therefore also for brittleness. One method of recovering fibres is so-called retting, which is used in the case of flax and hemp, for instance. In this ancient process, the vegetable stems are exposed to moisture by being left in the field or in retting pond. Bacteria then attack and break down the pectins which hold the individual fibres together. Mechanical methods are then applied for further processing of the stems prepared in this way. Figure 2

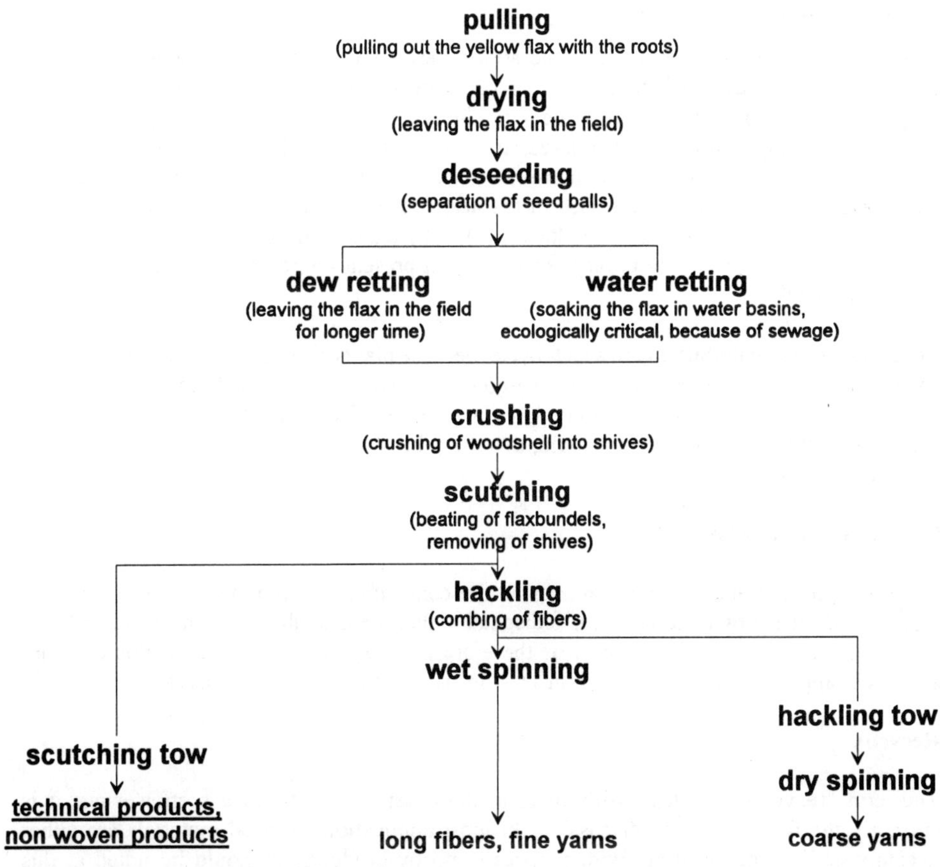

Figure 2. The classical production route for flax fibre

roughly outlines the classical fibre recovery cycle. Newer processes, such as the surfactant process, mechanical cottonizing and the steam-dissolving process [1], which we will not examine in more detail here, however, are nowadays used.

Availability

Due to the very diverse fields in which fibres are utilized, there is no reliable data on the quantities used. Nor, indeed, are there any worldwide producer associations (as there are, for instance, in the chemicals industry), with the result that an actual-situation analysis is difficult. Figure 3 does, however, attempt to outline the quantities produced in 1992. It impressively demonstrates that, even assuming the fulfillment of an optimistic trend (approx. 5 kg fibres per vehicle), natural fibres are, as a resource, practically inexhaustible.

Arguments for the use of regenerable raw materials

The CO_2 problem

In view of the increasing burden of the atmosphere's CO_2 content, particular importance attaches to the processes which occur in energy conversion, and to those which occur in the propulsion of a motor-vehicle in particular. The warnings uttered by climatic researchers, until very recently still lightly dismissed, are now beginning to influence the development of future engineering products ever more (particularly in automotive engineering). This new awareness will probably be expressed in the near future, as many politicians at least wish, in the form of ever more stringent laws (CO_2 tax) against the background of a holistic balance. Only the current temporary recession is at present retarding the imposition of this additional taxation.

Regenerable raw materials create a CO_2 cycle, because plants absorb approximately the same amount of CO_2 during their growth as a result of photosynthesis as is subsequently liberated upon combustion or decomposition. A (if we are realistic, small) contribution to the reduction of the greenhouse effect could, therefore, be made, since no additional CO_2 would enter the atmosphere.

Conservation of resources

A further positive and leading argument is the concomitant conservation of fossil energy sources and other raw materials. It is indisputable that the available amounts of natural gas, mineral oil and coal are finite and it is therefore urgently necessary, even if this does not at present appear obvious, to switch at least in part to regenerable resources.

Recycling

The term "recycling" is dealt with again in more detail in a separate section, since it is currently the focus of much discussion. In most applications, natural materials are either decomposable or combustible with practically no by-products. It should be noted at this point that the production of CO_2 in the utilization of organic by-products, residues and waste for energy purposes is, for reasons which cannot be rationally explained, not tolerated by the human population, in contrast to the production of CO_2 involved in composting. Large-scale educational work is still necessary in this field.

These remarks are, anyway, true only in cases in which thermoplastics, or even thermosetting resins do not prevent decomposition by "encapsulating" the organic fibres.

In such cases, other reprocessing methods are necessary, as discussed below.

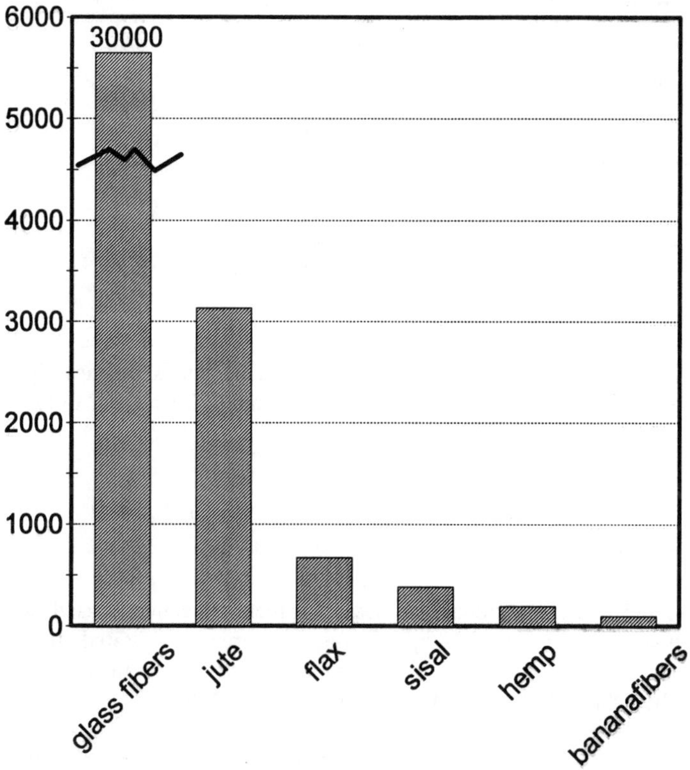

Figure 3. Amounts produced (natural fibres v. e-glass fibre); 1000 t/a. Source: e-glass, banana: [3] as per 1990. Natural fibres: [4] as per 1992

Alternatives for agriculture

The marketing of high-quality technical products could (even - or perhaps particularly - for European agriculture) be an interesting alternative to foodstuffs production. This might make it possible to bring heavily subsidized non-utilized land back into production, relieving the burden on the public budget. This would, however, presuppose modified knowledge and production methods in agriculture, for which, according to recent research, there is acceptance, at least in the southern federal states of Germany. Daimler-Benz Research and Mercedes-Benz have offered to take an active role in the overall chain of development.

Cost situation

Due to the diverse crop sources and greatly fluctuating demand, no stable price structure has yet become established.

tecnical glass fibers	3,0...4,0
reprocessed cotton	0,7
flax tow	0,3...0,8
nail flax (combed flax)	1,0...1,5
sisal	1,5...2,0
jute	1,0...1,3
banana fibers	1,4...1,8
coconut fibers	0,7...0,9

Figure 4. Prices for natural fibres [DM/kg], mid-1993, [1], [5]

The example of flax can be used to illustrate the way individual fractions are heavily dependent on one another as coupled products in each case. The market for natural fibres can be subdivided into three price and quality categories:

- Fibre waste and secondary fibres (short or milled fibres)
- Long fibres
- High-performance engineering fibres

For picked flax (linked product in the traditional flax beater), which is frequently used to make fleeces, prices of between 0.3 and 0.8 DM/kg (with a rising trend) are stated. If it proves possible to enter higher-value applications with long fibres, however, an interesting economic advantage could be achieved at prices of between 2 and 3 DM/kg.

Technical properties

The principal emphasis in efforts to increase the use of natural fibres is found in the reinforcement of various plastic matrices.

The disadvantages of natural fibres

Natural fibres are subject to severe fluctuations of quantity and quality. This problem can, however, be countered by diversifying the planting regions and blending of various qualities, thus achieving standardization. A further problem may be associated with one of the product's recycling advantages, i.e., uncontrolled decomposition of the constituents of natural materials, which, logically enough, may only be permitted once the period of technical utility has finished. Successful additive systems have already been developed for open-pored systems (the only case in which the danger occurs), however.

A further aspect is found in PR and marketing. The positive advertising anticipated by most suppliers for the use of natural fibres appears always to have a "cheap" image attached to it by the clientele, particularly in the high-tech and, therefore, high-price segment. Further developments are anticipated in this area.

Technical advantages of natural fibres

In addition to the above-mentioned negative criteria, however, there are a whole series of technical advantages. Due to the low density of natural fibres, usually of less than 1.4 g/cm^3, i.e., of a figure comparable to the density of aramide fibres, good specific strength properties can be achieved, and in the case of flax fibres from special breeds attain the levels reached by glass fibres.

	density (average) [g/cm3]	specific tensile strenght [MPa * cm3/g]	specific modulus of elasticity [GPa * cm3/g]
tecnical glass	2,55	705...1098	28...29
flax	1,5	195...700	9...20
sisal fibers	1,45	586	11...26
jute fibers	1,5	213	13...23
banana fibers	1,35	392...677	20...24

Figure 5. Comparison of natural fibre and glass fibre properties

It becomes apparent on a general overview that precise knowledge of functional loads will make it possible for material-oriented design to achieve component-weight reductions. This effect can be even further enhanced with the use of long-fibre reinforcement in foam structures.

A further significant advantage, given appropriate manufacture technology, is the good bending flexibility of natural fibres as compared to corresponding conventional materials. These properties could be of particular advantage in special applications such as door internal trims, in which the special benefit could take the form of improved crash behaviour. Natural-fibre-reinforced components do not splinter, they fail with a characteristic tear. Penetration tests performed on various materials used or planned for use in door trims confirm these properties.

In contrast to synthetic fibres, natural fibres are already composite materials consisting of cellulose, lignin and small amounts of other substances. This lignin acts in certain processes, and particularly in thermoset systems as a binder and thus permits a reduction in the use of conventional binders such as phenolic resin (PF) and epoxy resin (EP), to figures below 20 %. It is also possible to add further lignin (obtained, for instance, as a by-product in paper manufacturing) to the thermoset binders, in addition to the natural lignin contained in the fibres and to thus reduce the synthetic binder content even further, or even to bind the fibres entirely with lignin or starch and thus create a genuine natural composite material. Results of tests on materials consisting of fibre fleeces bound solely with lignin are already available, although they are not yet entirely satisfactory.

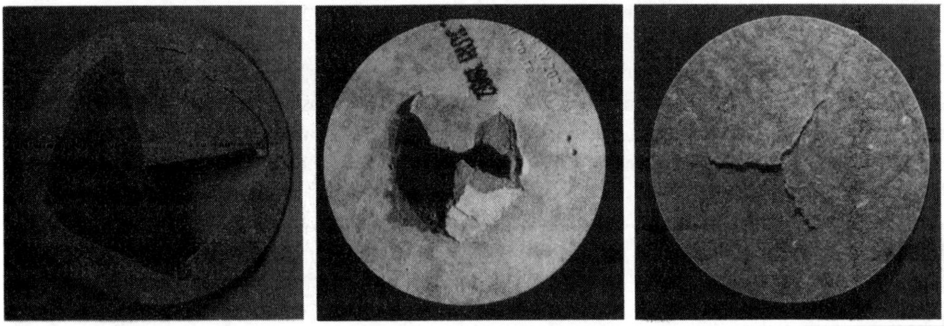

Figure 6. Penetration tests performed on materials for interior door trims

In practically all production processes involving glass fibres, the abrasiveness of this material is a problem, particularly in high-volume production runs. The abrasiveness-critical shear edge in GMT (glass-mat-reinforced thermoplastics) production may be mentioned as an example. The complaints made by assembly staff regarding their exposure to airborne glass-fibres from cut edges have increased significantly recently. Present-day knowledge indicates that there is probably no toxicological danger, but there is a severe nuisance, nonetheless. Such risks and disadvantages are unknown up to now with natural fibres.

The tubular structure of most natural fibres gives them interesting acoustic and thermal insulation properties. Investigations into this field had not been completed by the submission deadline for this article, however.

Misting-up of screens and removal of moisture and humidity around seats can produce problems in the interiors of vehicles and result in reductions in safety and comfort. Sophisticated use of natural fibres could exploit their local climate modifying properties in this context. The same function can, of course, also be performed by an expensive and complex air-conditioning system with an interior dehumidification function.

Manufacturing processes

The type of matrix (thermoset or thermoplastic) determines the selection of the production process. Both pressure moulding process and injection-moulding can be used. The thermally limited load capacity of natural fibres and the need to preserve fibre length prohibit multistage processes and long residence times. Fogging, odour problems and the fibres' water balance necessitate extremely careful process design.

The film-stacking method, the extrusion process, the Triform process and the textile process are all used in pressure moulding of thermoplastic matrices, the manufacturers active in these fields all operating their own refinements and development versions.

In the film-stacking method, fibre fleece and thermoplastic sheet are stacked alternately in layers, then heated, placed in a mould, and then pressure moulded. The thermoplastic melts during the heating process and impregnates the layers of fleece.

In the extrusion press process (also known as the EXPRESS process), which is under further development at Daimler-Benz Research in Ulm, the thermoplastic is extruded into the open mould using a fish-tail die, and then pressure moulded.

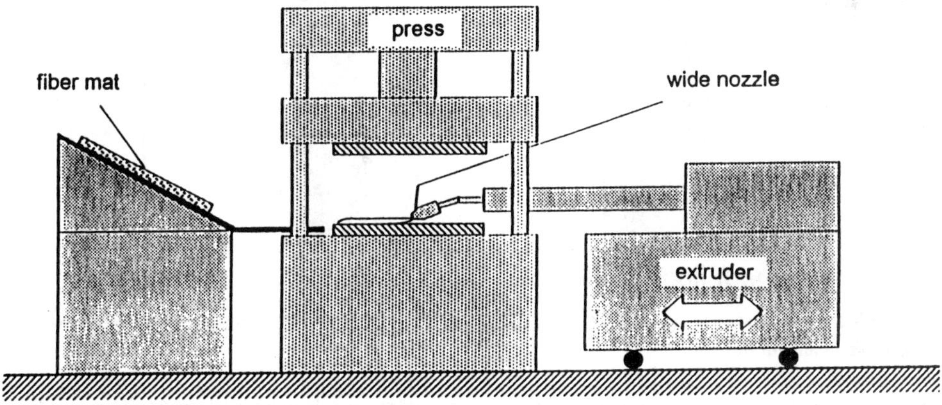

Figure 7 The extrusion press process, [6]

The particular advantage of this process can be found in the avoidance of semi-finished products and the non-destructive application of heat. Where polypropylene (PP) is selected as the matrix, fibre/matrix adhesion must be enhanced by means of suitable site treatments. The use of adhesion-promoting agents based on maleic acid anhydride makes it possible to optimize the intermaterial adhesion between hydrophilic fibres and hydrophobic matrix.

In the Triform process, as used increasingly in Brazil, for instance, various fibre materials, such as jute sacks or textile waste, are shredded and processed. Old thermoplastic sacks are also hacked up or shredded. These two fractions are then intimately mixed, shaped to form semi-finished products, heated and pressure moulded. A similar process is practiced in Germany using textile waste and a thermoset binder (Triflex). In order to achieve products with the most consistent possible quality, largely single-category (or even single-type) thermoplastic material must be used. Only a restricted output level can be anticipated, depending on the age of the jute starting materials and the nature of their storage. This material has, up to now, been used only in interior trim elements.

In the textile process, a semi-finished product is formed from natural and thermoplastic fibres by means of carding and needle punching, then heated, placed in a mould and pressure moulded. It is also possible to produce from natural and thermoplastic fibres, by means of spinning, for instance, a material which can also be shaped into highly varying forms by means of heating and pressure moulding.

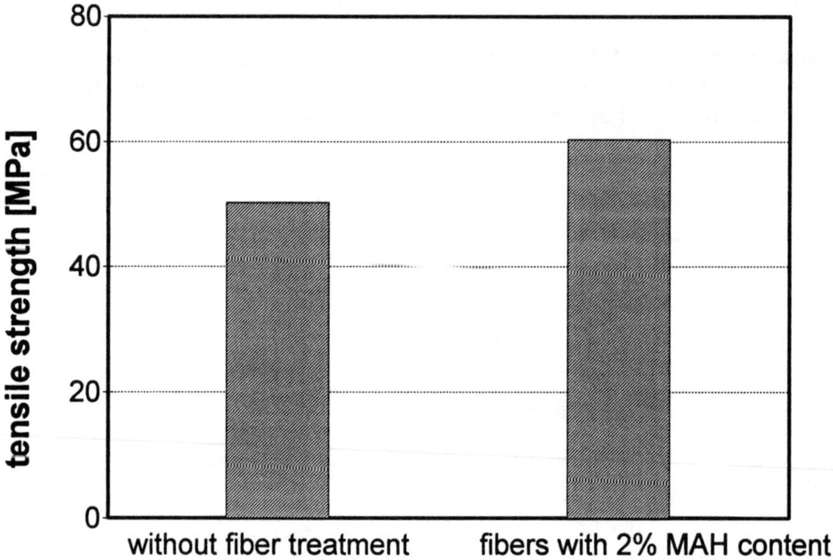

Figure 8. The effects of sizing to improve fibre/matrix adhesion (flax fleece/PP, film-stacking process; MAH = maleic acid anhydride)

Other pressure processes employing thermoset matrix materials are also known. They differ from one another, above all, in the method of introduction of the binder. It is possible, for instance, to dissolve the binder (in cyclohexane, for instance), to impregnate the fibre fleece with it, then dry the fleece, resulting in the binder remaining in the fleece. The fleece treated in this way is then pressure moulded. Another possibility is to scatter the binder in powder form onto the fleece and then to pressure mould the fleece. As well as the pressure moulding process, the injection-moulding process has already been mentioned. Here, granulate produced by the thermoplastic matrix processes mentioned is used. The advantages of the long-fibre properties can then, however, no longer be fully exploited.

Applications in commercial vehicles

Application potentials involving the use of natural materials are outlined in an overview (Figure 9).

At the "ECO 92" exhibition in Rio de Janeiro, which was held against the background of the major global environmental conference, Daimler-Benz presented, in the form of a demonstration model, work performed by Mercedes-Benz do Brazil in the field of natural materials. Prototype components produced from natural materials were exhibited. The range extended from door handles, large door, back panel and roof trim elements, via upholstery elements such as seats, headrests and anti-dazzle visors, through to heavily load-exposed components for the exterior, such as noise-suppression enclosures, spoilers, brake pipes and components consisting of natural rubber.

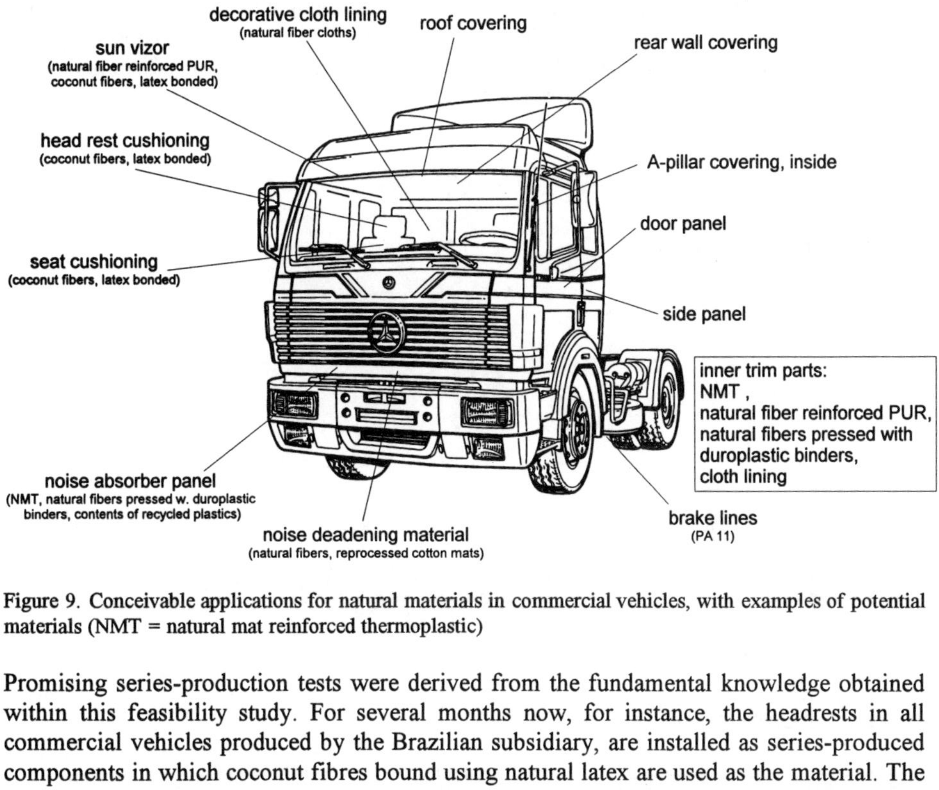

Figure 9. Conceivable applications for natural materials in commercial vehicles, with examples of potential materials (NMT = natural mat reinforced thermoplastic)

Promising series-production tests were derived from the fundamental knowledge obtained within this feasibility study. For several months now, for instance, the headrests in all commercial vehicles produced by the Brazilian subsidiary, are installed as series-produced components in which coconut fibres bound using natural latex are used as the material. The question of the extension of the knowledge obtained to other part-areas of the seat system is currently under examination. Initial results indicate that seat-cushion covers consisting of coconut fibre achieve a significant enhancement in comfort, particularly for long-distance buses. The use of this material is also conceivable in lightly loaded seat backs, but has been implemented up to now only in prototypes.

Driver's cab internal trim elements, such as those for the rear panel, A pillar, and roof, and for the doors in some cases, are in Triform (thermoplastic bound jute) and have been in series production for some time now. Sisal-fibre-reinforced PUR foam is also already in series-production use in these applications.

Testing of flax-fibre-reinforced door and side trim elements is being conducted in our heavy-goods vehicle category in Germany, on the basis of the knowledge obtained in Brazil. At present, EP is still used as the matrix, but is to be replaced, inter alia for recycling reasons, by a PP matrix.

In Germany, interior trim elements and the seats have first priority in terms of the use of natural fibres. These components are followed by shelves and pockets, the anti-dazzle visor, and the central console. A further potential for future applications can be found in noise-suppression enclosures. Development studies which are examining backing materials similar to GMT but consisting of flax-mat-reinforced PP and insulation materials consisting of reprocessed cotton fleece and/or flax fleeces are currently being conducted. Wheel housing covers, possibly in combination with recycled plastics, are also of interest.

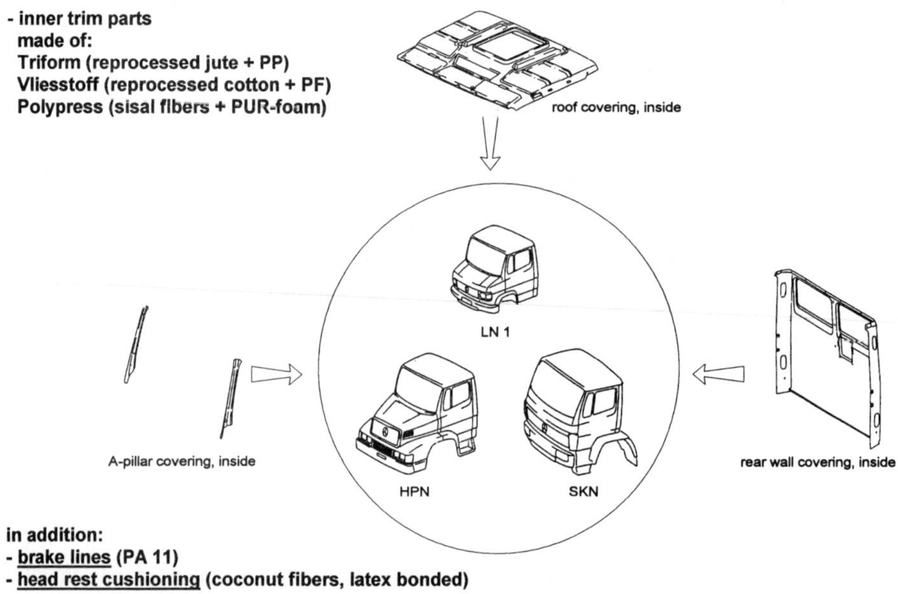

Figure 10. Series-produced components incorporating natural materials made at Mercedes Benz do Brasil

In the near future, a holistic balance will indicate whether it is worthwhile using natural materials to a greater extent in applications in which at present no technical or economic advantage appears achievable.

Motor-car applications

In principle, similar decision criteria apply in the car with regard to the use or non-use of natural fibres. In general, higher standards are set regarding surface qualities, and lower standards regarding misuse exposure. In most cases, the support components are laminated on, with the result that the advantageous technical properties predominate over the in some cases inferior visual characteristics. These technical advantages consist, above all, of weight reductions, and an extremely good penetration behaviour. The first component employing flax fibres to go into series production was the roof frame of the Mercedes 190, production of which terminated in July, 1993 (Figure 11). Door trim mountings of the same material are used in the follow-up model (C-class). This material is "Flexiform", consisting of 66 % flax plus reprocessed sisal and 34% expanded EP. All the tests performed have confirmed suitability for series production use.

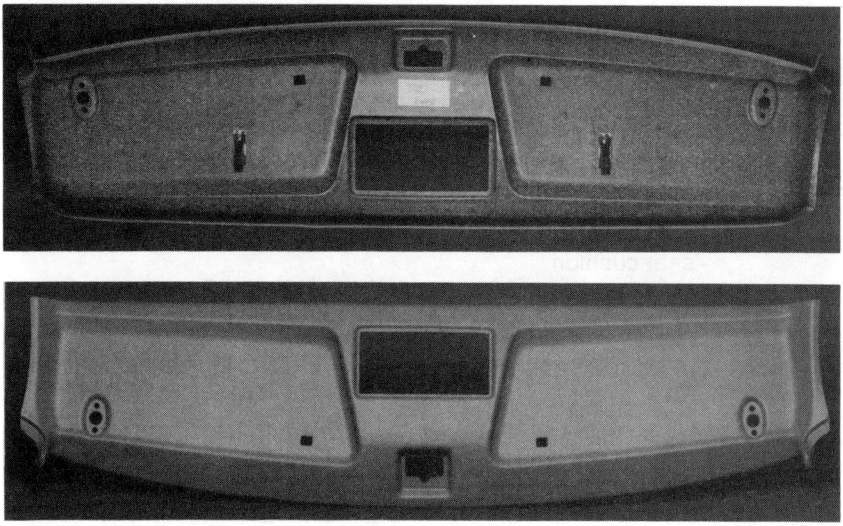

Figure 11. Roof framework lining for the model 190 series, consisting of EP-bound flax and sisal

Companies such as EMPE and Mapotex/Happich are pursuing diverse development projects in this field, although it is not always clear whether the price pressure exerted by the customer or new technical requirements are really the driving force behind the rapid process-engineering refinement involved. A further series-produced component using flax fibres is, for instance, the roof lining used in the vehicles produced by a competing company in southern Germany.

In addition to the interior trim elements already under examination, such as door trims, pillar trims, rear shelves and the seats, the boot (trunk) and exterior trim elements are also coming under increasing study. A spare-wheel cover consisting of (possibly starch bound) natural fibre fleece, and wheel housing covers in conjunction with recycled plastics, are, for instance, already under examination. Other potential applications are outlined in the overview.

It is apparent that holistic natural material systems with economic and ecological advantages and thorough functionality will be available in the not too distant future for motor-vehicle interior components.

Natural wood fibre and wood dust materials have, up to now, been left out of all these observations. This standard material has been a tried and proven extender for trim mounting elements for many years. Due to its weight and brittleness problems, however, this material has come into disrepute when compared to the bast fibre materials, and there will probably be no expansion in its range of uses in future.

door covering

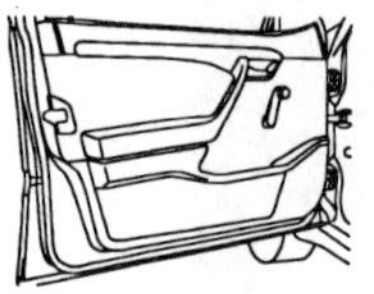

- panel: wood particelboard, natural fibers pressed with duroplastic binders, NMT, cloth lining
- wood trim: NMT-holder, wood veneer

front seats

- head rest
- back rest
- seat cushion
  (coconut fibers, latex bonded, with natural fiber cloth lining)

- seat covering (backside): NMT

roof liner

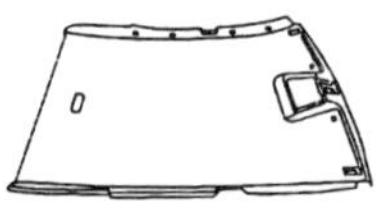

- natural fiber/EP, PUR, cloth lining compound
- design with roof frame covering (W201):
  Flexiform (flax/sisal, EP)

Figure 12. Applications for natural materials in cars, components at a high development level

A frequent objection to the use of natural fibres is the claim that dimensional tolerances cannot be met, due to the fibre material's high water absorption. Measurements made on the roof frame and pillars impressively document that all changes in dimension, even when the component is subject to load, remain within the permissible tolerance. Even starch and lignin bound natural fibre fleeces exhibit extremely promising properties vis-à-vis this problem. In tests on phenolic resin systems in which approx. 50 % of the resin was replaced with lignin, no property losses were observable and the modulus of elasticity in flexure was actually increased [7].

Recycling and environmental safety

The fact that a motor-vehicle manufacturer operates at the very cutting-edge of progress is illustrated initially by the large number of criteria which have priority at Mercedes-Benz, for instance. In addition to the demands for active and passive safety, for cost-effective manufacturing and operation, for reliability throughout the product as a whole, there is, today more than ever, a concern for environmentally safe products. Recycling and ecological

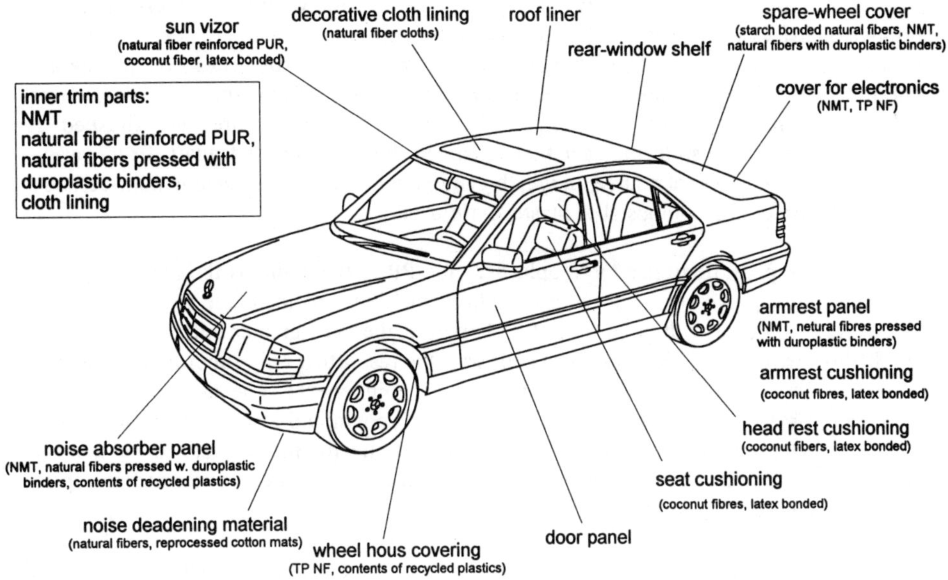

Figure 13. Conceivable applications for natural materials in cars (NMT = natural fibre mat reinforced thermoplastic, TNFP = natural fibre reinforced thermoplastic)

consciousness must not only play a role during the design of vehicles but must also extend as an overall function throughout production, service and disposal. Since environmental safety and recyclability are now important components in every development specification, the result of a holistic evaluation of regenerable raw materials, and of natural fibres, in particular, will need to be increasingly incorporated. Paramount significance attaches to the search for <u>optimum</u> recycling. Efficient and economically viable solutions must be found. There are certain criteria which must be fulfilled for rational materials recycling:

- Components must be marked
- Materials should not be significantly aged after they have been in use
- The reprocessing (recycling) costs should not exceed the value of the material
- Multi-material assemblies should be separable into single-type units (or should be rationally reutilizable as a mixed unit
- Component size and material mixture must permit cost-effective removal and dismantling.

The recycling technology used or planned at Mercedes-Benz provides for the prior removal and dismantling of large and readily accessible components before the residual carcass is submitted to metallurgical recycling. All the components incorporating natural fibres which have been mentioned here, whether already implemented or only conceived, fulfil these basic prerequisites.

Natural fibres are in principle always biologically degradable, whereas the thermoplastic or thermoset matrices which surround them, at least as currently known, are not. And they are not degradable, even when they are not based on mineral oil but instead on regenerable natural oils, like PA11, or the cellulose plastics, for instance. The claim that natural-fibre-reinforced components can be composted when their useful life has finished is, unfortunately, false, despite the fact that it can be heard again and again from various quarters, except in cases, of course, in which genuinely biological degradable matrices, such as starch, lignin or the compostable plastics, are used.

Should it prove possible to use thermoplastics, and PP in the main, as the matrix, then the known recycling processes, such as granulation, melting and injection moulding can be utilized. As is the case with glass fibre, too, however, the even greater initial lengths of the natural fibres will be significantly shortened, with the result that only subordinate "cascade" applications will be possible in injection-moulded components. The use of a thermoset matrix would make possible particle recycling, as is known and also practiced in the case of SMC (sheet moulding compound). This will continue for many years to be the only materials recycling alternative, since this combination will continue to occur in life-expired vehicles for the next ten years.

Where natural binders are used, there is a possibility of long-term compostability which will not, however, signify a relaxation of the need for landfill capacity in the short term. Additions, such as site treatments, to improve fibre/matrix adhesion, as well as the necessary flame-retardants, again impair such complete degradability, but are indispensable for the achievement of the required property profiles.

It should here again be stated unequivocally that thermal recycling is technically, ecologically and in energy-balance terms suitable for all the systems discussed and should be preferred in the majority of cases. If the components conceived in the future contain significantly more than 50 % regenerable raw materials, the argument of "conservation of resources" will retreat as an impulse for the creation of materials cycles.

Prospects

From the enormous group of natural materials available, the above deliberations have focussed on only a small selection of the regenerable reinforcement fibres for plastic components. Although cellulose fibres (in glove-compartment linings, for instance), cotton (in trim elements) and wood fibres (in door trims, for example) are already used in vehicle engineering by many manufacturers, but generally with a subordinate engineering reinforcement function, the real advance in natural technological fibres has only just begun.

The following clear trends are discernible:

- Thermoset resin systems are gradually losing their technical and ecological justification, at least in temperature ranges below 100 °C.

- Single-substance systems (with PP particularly important) will begin to predominate
- Assuming appropriate breeding and further developments, both tropical and domestic fibre materials will be able to achieve the high level of performance offered by glass fibres, and will no longer be used solely as extender materials.
- The principal applications will be restricted to flax, sisal, jute, coconut and banana fibre, in that order
- In the near future, natural fibres will increasingly be the material of choice, and not only for cost reasons but instead thanks to weight and functional advantages. Given the increasingly long periods which will with certainty be spent by human beings in motor-vehicles in future, it is necessary to make vehicle interiors as environmentally friendly, and as free from odour and fogging as possible, to equip them with a pleasant tactile ambience and to further optimize the passive safety of collision-endangered zones. Nature will be able to make a large contribution to all this. Everyone everywhere is making a late start, so that there is all the greater need to work the more intensively on these problems.

The Mercedes-Benz motor-car development division and the Production Research and Environmental Systems division of Daimler-Benz AG have made a significant joint contribution to the promotion of the use of natural fibres; intensive assistance was also provided by the cooperant suppliers selected, EMPE and Mapotex/Happich.

References

1 Tubach, M., Kessler, W.: Institut für Angewandte Forschung, Reutlingen: Marktsituation und Einsatzbereiche von Flachs, p. 4
2 Renz-Rathfelder, S.: Faserpflanzen, Stadt Frankfurt am Main, Dezernat für Umwelt, Energie und Brandschutz, Palmengarten, Sonderheft 18 p. 28-34
3 Baumgärtel, H.: BASF/EKT, Ludwigshafen, Sisalmattenverstärktes Polypropylen
4 FAO production yearbook, Vol. 46, pp. 194-199
5 Transferzentrum für nachwachsende Faserrohstoffe Reutlingen: Trägerteile für LKW aus Flachs, Auftraggeber: Mapotex
6 Fölster, Th.: Daimler-Benz Research, Natural Materials for Composites', Technical Report
7 Transferzentrum für nachwachsende Faserrohstoffe Reutlingen: Automobil-Innenteile aus Flachs: Neue Bindemittel

Plastics applications to meet safety requirements

Winfried Köppen

Introduction

An analysis of all accident incidents on the basis of the consequential costs of injury indicates that side impacts make up more than 45 % of the total number of collisions. In Europe and the USA, this circumstance is recognized in the form of corresponding side-on crash legislation, but with the use in each case of differing test configurations and dummies. The introduction of these laws is planned as follows:

Europe: Probable introduction in October, 1995, for all newly registered vehicles

USA: To be introduced on September 1, 1994 for all of a manufacturer's production for the US market

or, as an alternative: Graduated introduction, for 10 % of production by September 1, 1993, 25 % by September 1, 1994, 40 % by September 1, 1995 and 100 % by September 1, 1996.

The test configurations are shown in Figures 1 and 2. The diagonal impact of the US barrier is intended to represent the primary motion of the collision vehicle at 24 km/h. In addition, the barriers have deformation elements of differing geometries and rigidities.

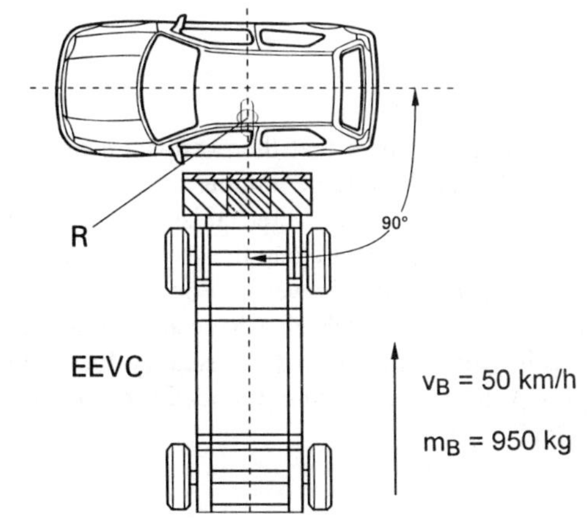

Figure 1 European side-on crash test configuration (EEVC proposal)

Both sets of legislation pursue the objective of controlling the acceleration acting on the occupant when a side-on collision occurs, wherever possible in such a way that he has a chance of surviving. The particular legislation either specifies permissible maximum deformations and forces or maximum permissible accelerations acting on the dummies, which are also specified in each case. As a result, the deformable zone between the interior trim and the outer skin of the vehicle must be tailored in terms of its deformation-force level to the dummy and, therefore, to a human being.

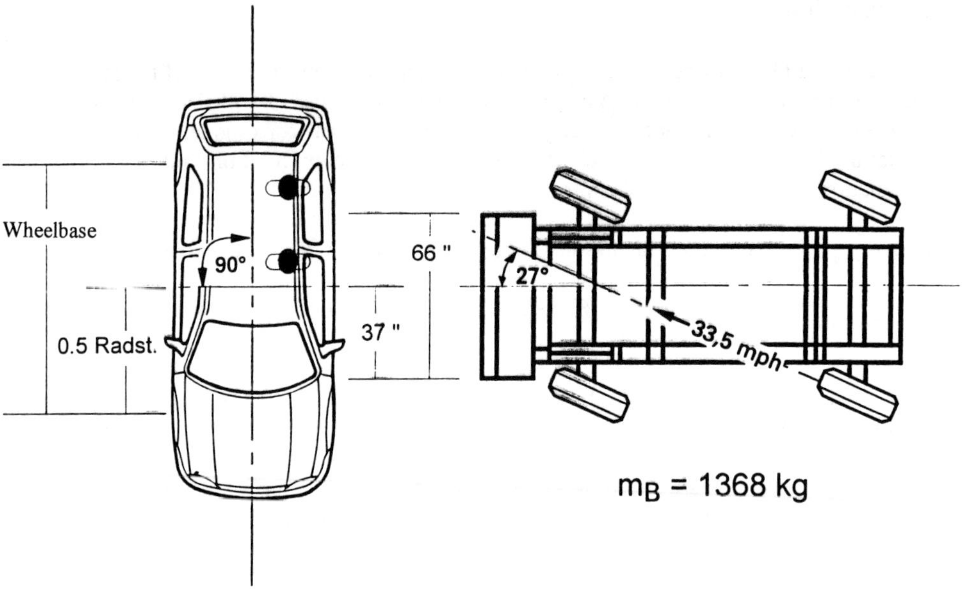

Figure 2 US side-on crash test configuration (FMVSS 214)

This fact becomes particularly significant when one remembers that the occupant's impact zone in a side-on collision is taken up in a present-day car by a large number of attachments and internals, such as pockets, loudspeakers, window winder mechanisms, etc. The development engineer is therefore forced to accelerate the occupant along the shortest possible path, i.e., at a high level of force, without exceeding the limits, however. For this purpose, Volkswagen is increasingly employing foam-type plastics, which also have the additional function of protecting the occupant against locally concentrated loads caused by localized intrusions into the passenger compartment. Figure 3 shows the padding provided for a side-on collision, using the example of the two-door Golf.

The side-on impact process is utilized in the following discussion to represent both crash procedures, using the example of the US side-on collision. Figure 4 shows the situation which occurs at the precise time of impact of the barrier with the vehicle. The barrier contour (1) with its hard "bumper zone" (dark) can be discerned on the extreme left. The door with its attachments, accessories and internals, side protection beam (2), padding (3) and door interior trim (4) is shown in cross-section. The positioning of the seat and the dummy are identical to those in a real vehicle. Half of the jacket has been removed, to improve comprehension.

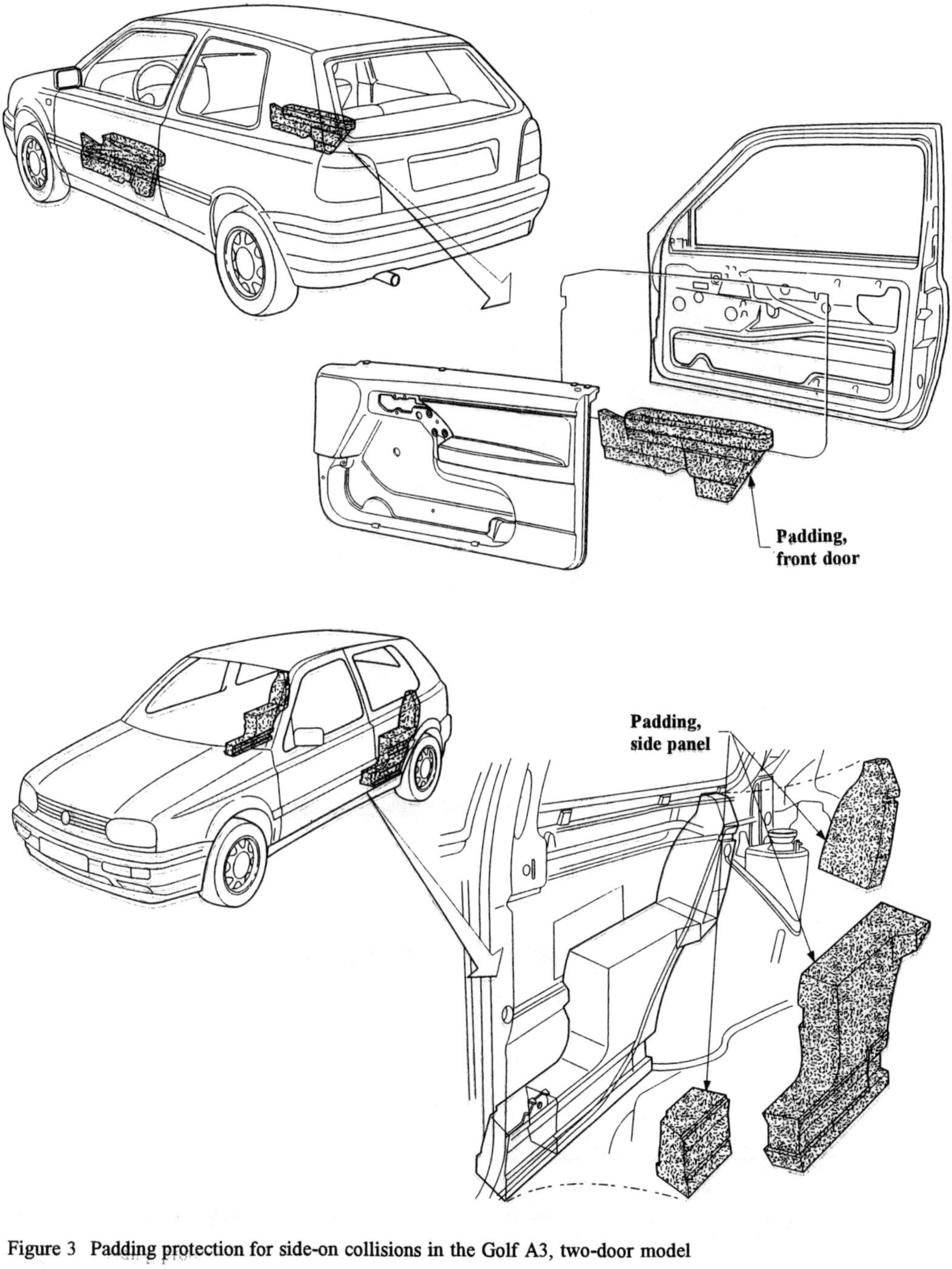

Figure 3 Padding protection for side-on collisions in the Golf A3, two-door model

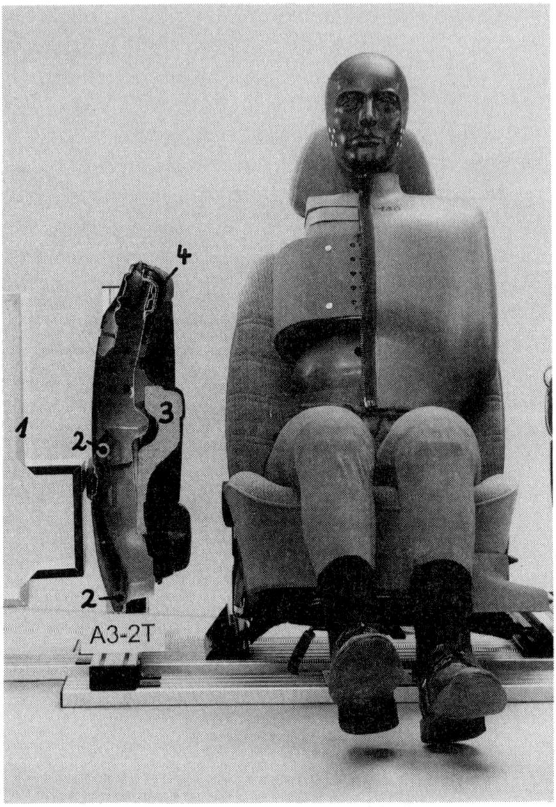

Figure 4 Positioning of the front seat dummy as per US-FMVSS 214 relative to the door, its internal components and the barrier in the Golf A3, two-door model and convertible

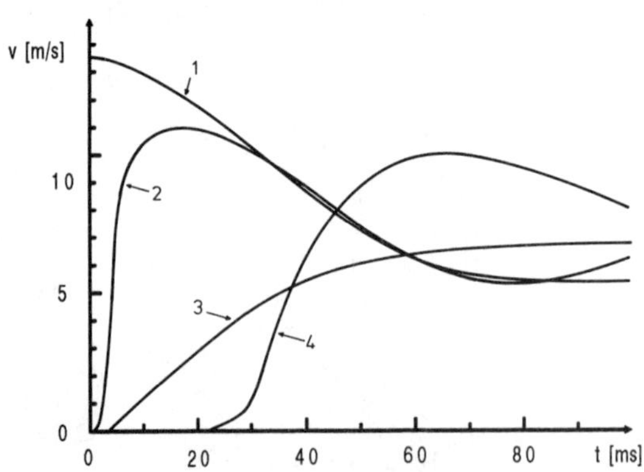

Figure 5 Velocity/time diagram for the transverse to vehicle direction in a US side-on collision (1. Barrier, 2. Side structure, 3. Vehicle COG, 4. Occupant)

Figure 5 shows a typical velocity/time (v/t) diagram for the transverse-to-vehicle direction of a US side-on collision. At time t = 0, the barrier strikes the vehicle and penetrates into it in this direction at a velocity of 13.5 m/s (Curve 1). The measuring apparatus fitted in the vehicle and the dummy do not yet register any reaction. The first measuring point to react is then the side structure of the vehicle facing the impact, represented by the "B pillar/left" measuring point, which is accelerated to approximately the velocity of the barrier (Curve 2). The acceleration of the center of gravity of the vehicle (center tunnel/mid-point Curve 3) occurs more moderately and somewhat later. The difference between these two velocities expresses the vehicle's rate of deformation.

The occupant is struck by the door or side trim after about 20 to 30 m/s, depending on the particular vehicle (Curve 4). Under the deformation of the dummy, on the one hand, and the cavity between the trim and the outer door skin, on the other, the occupant is accelerated to the velocity of the penetrating structure. Certain limits must not be exceeded during this process, which is essentially completed after around 50 m/s (see Figure 6).

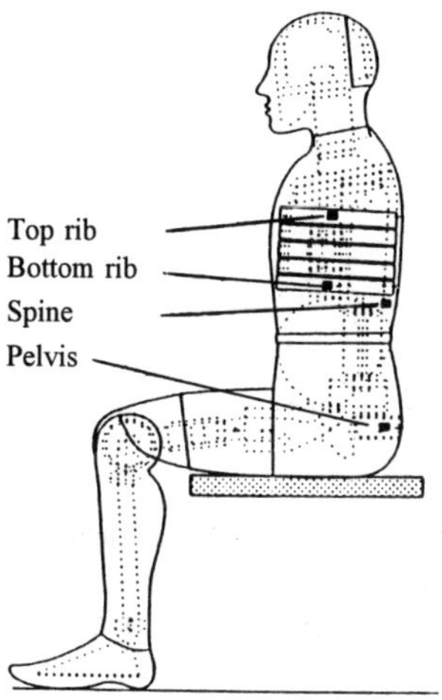

Thorax:

$$TTI\,(d) = \frac{A_R + A_W}{2}$$

$\leq$ 85 g with 4 doors

$\leq$ 90 g with 2 doors

A_R Maximum ribcage acceleration in g

A_W Maximum acceleration of spine in g

Pelvis:

$A_B \leq 130$ g

A_B Maximum acceleration of pelvis in g

Figure 6 Legal requirements of US Standard 214

Theoretical principles:

The following simplifying assumptions can be used to derive a simple and plausible formula for the deformation spaces and padding thicknesses necessary in a side-on collision:

1. The change in velocity of the penetrating door and side-panel structure during the dummy's acceleration phase is ignored. After the first dummy/door (side-panel) trim

contact, it approaches the occupant at a constant velocity v_T, i.e.: in time t, the side structure travels

$$x = v_T \cdot t \tag{1}$$

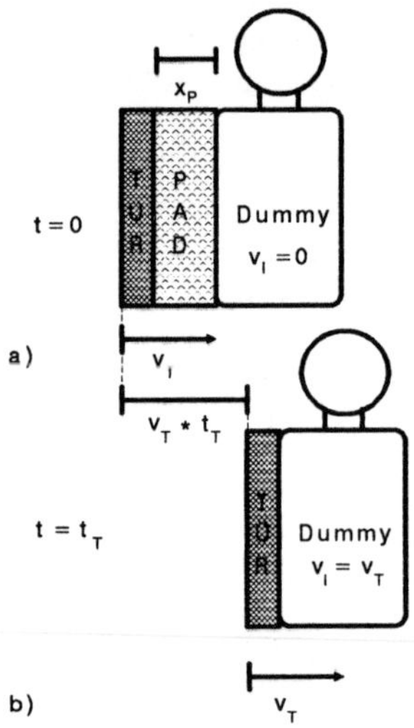

Figure 7 Effects of padding in a side-on collision:
a: Door and padding strike the motionless dummy at velocity v_T
b: Dummy reaches the velocity of the door; the thickness of the padding has been used up

2. A layer of padding of thickness x_P (see Figure 7a) is present between the non-deformable door structure and the occupant upon his initial contact with the door interior trim and accelerates the occupant to velocity v_T at time t_T with a constant acceleration of a_I, i.e.:

$$v_I = v_T = a_I \cdot t_T \tag{2}$$

At this point, the occupant has travelled

$$x_I = 0.5 \cdot a_I \cdot t_T^2 = 0.5 \cdot v_T^2/a_I \tag{3}$$

In addition, with ideal padding design, the padding thickness should at time t_T have been "consumed" for acceleration of the occupant, i.e., the non-deformable door structure has "caught up" with the occupant (see Figure 7b):

$$\begin{aligned} x_T &= v_T \cdot t_T = v_T^2/a_I \\ &= x_P + x_I = x_P + 0.5 \cdot v_T^2/a_I \end{aligned} \tag{4}$$

The required padding thickness is therefore

$$x_P = 0.5 \cdot v_T^2/a_I \tag{5}$$

The same result is obtained if a body with kinetic energy E_{kin} is decelerated at constant deceleration force $F = m \cdot a$ along path x, i.e.:

$$E_{kin} = 0.5 \cdot m \cdot v^2 = E_{Br} = \int F\, ds = m \cdot a \cdot x \tag{6}$$

Equation (5) can be used to directly estimate the minimum necessary deformation depth for doors and side structures for a particular maximum permissible occupant acceleration. In practice, however, this deformation zone must be greater, since the accelerative force does not assume a constant level. The maximum figure for accelerative force, on the other hand, is decisive for fulfillment or nonfulfillment of the side-on collision law.

For this reason, the efficiency of an energy absorption system is defined as follows:

$$\eta^* = \int F\, ds/(F_{max} \cdot x_{max}) < 1 \tag{7}$$

In practice, the padding materials can be deformed only up to their densification thickness; i.e., total padding thickness cannot be used for deformation. The padding thicknesses necessary therefore increase vis-à-vis Equation (5) by a factor of $1/\eta$:

$$x_P = (0.5/g) \cdot (v_T^2/a_I) \tag{8}$$

or

$$a_I = (0.5 \cdot v_T^2) / (\eta \cdot x_p) < a_{I,limit} \tag{9}$$

in which

$$\eta = \int F\, ds/F_{max} \cdot x_P < \eta^* < 1 \tag{10}$$

In Equation (9) the two basic strategies for fulfillment of the side-on collision laws are clearly expressed:

1. Only moderate acceleration acting of the occupant achieved by means of force and acceleration limiting deformation zones of adequate thickness x_P in the area of impact with the occupant:

A high efficiency η, i.e., the most rectangular possible force-deflection characteristic of the padding material and a low solid compaction thickness, is necessary in order to make optimum use of the limited spaces available in the door for acceleration of the occupant during a side-on collision. This strategy is rational only if the survival space is adequate and the side-structure rigidity correspondingly high; for only in this way can it be assured that the occupant will not be injured by excessive vehicle deformation, on the one hand, and that, on the other hand, he can easily be extracted from the vehicle.

An increase in side-structure rigidity can be achieved both by means of structural provisions in the form of reinforcements and by means of structural foam components which achieve a systematic modification of the structure's deformation behaviour. In addition, it produces a:

2. Reduction of the velocity of door/side-structure penetration v_T in a dummy crash.

v_T, however, has a lower physical limit, which is defined essentially by the ratio of the masses of the two colliding bodies. For this reason, an increase in the rigidity of the side-structure and the concomitant reduction in the rate of penetration are not the only way of fulfilling the requirements of the side-on collision laws and must therefore be combined with the "initial strategy", i.e., with padding designed for the particular vehicle. A test method for the selection of side-on collision padding materials, complete with corresponding measured data, is presented below:

The test method at VW

The limits mentioned above for occupant accelerations correspond to maximum surface pressure figures acting on the corresponding regions of the dummy's body. Calculation data indicate that figures of approx. 0.11 N/mm^2 in the chest area and of approx. 0.45 N/mm^2 in the pelvic area must not be exceeded.

Figure 8 Test stand for selection of padding materials

Since a dynamic test more closely simulates behaviour under crash conditions than a quasi-static test, VW has developed a method for selection of materials with the above-mentioned specifications. In this, anatomical blocks representing the dummy's rib and pelvic areas are retarded at defined weights and rates of deceleration by the material under test. Figure 8 shows the test apparatus with the material under test (at the bottom), onto which the body block (at top) can be dropped from a defined height. In combination with mathematical evaluation methods, acceleration measurements supply the force-deflection characteristics curves and the corresponding energy absorptions.

Test results

The measured data presented here was obtained with a drop weight of 33.45 kg and a velocity of approx. 30 km/h, equating to a kinetic energy of 1.15 kJ. In a real side-on collision, this energy would have to be absorbed approximately in the pelvic area. Since the energy to be absorbed by the padding is lower in the chest area, these measurements can also apply in energy terms to this area of the body, too.

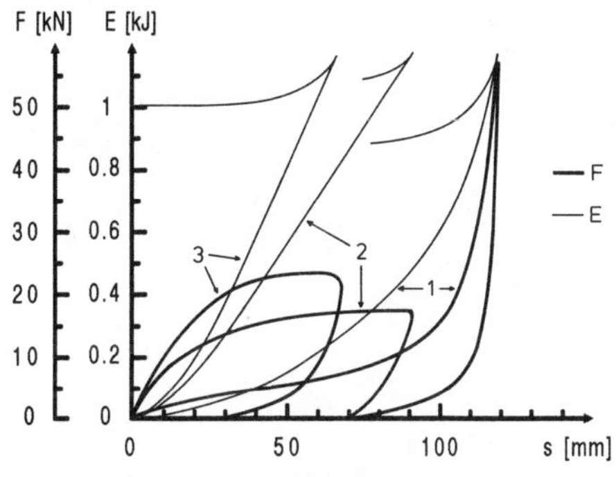

Figure 9 Force/deflection characteristics showing appurtenant energy absorption plots from drop tests using a thorax anatomical block (33.45 kg, v_0 = 30 km/h) for three different densities of PUR foam (thickness 130 mm): 1: 40 g/l, 2: 80 g/l, 3: 100 g/l

Figure 9 shows three force-deflection characteristic plots obtained for PUR foams in the manner described above, and their energy-distance plots for a foam thickness of 130 mm. The 40 g/l foam in Curve 1 is too soft for the energy of 1.15 kJ to be absorbed; after an initially gentle rise in force to approx. 10 kN, the foam reaches solid compaction at a maximum force of 57 kN. In Curve 2, the foam hardness was selected in such a way that the energy is completely absorbed at precisely the point at which the foam has been deformed practically to its solid compaction thickness. The foam's properties are optimally exploited under these circumstances. In the case of the even harder foam in Curve 3, on the other hand, the energy is absorbed significantly earlier at a higher force level.

Efficiency η^* of energy absorption in accordance with Equation (7) depends, on the one hand, on the material and also, however, on the geometry of the anatomical block. Under the given circumstances, the load-bearing surface of the foam becomes greater as depth of penetration of the anatomical block into the material increases. As a result of this geometrical effect, PUR foams, for instance, exhibit an efficiency of 0.5 at low foam deformation levels, which may rise to a maximum of 0.75 shortly before the material reaches solid compaction at greater foam thicknesses and the then possible greater deformation depths. In the case of PP, the ratios are approx. 10 % poorer. In the solid compaction range, which starts at approx. 60 to 75 % compression, depending on material density, η^* decreases significantly, a circumstance for which allowance must be made when specifying padding thickness.

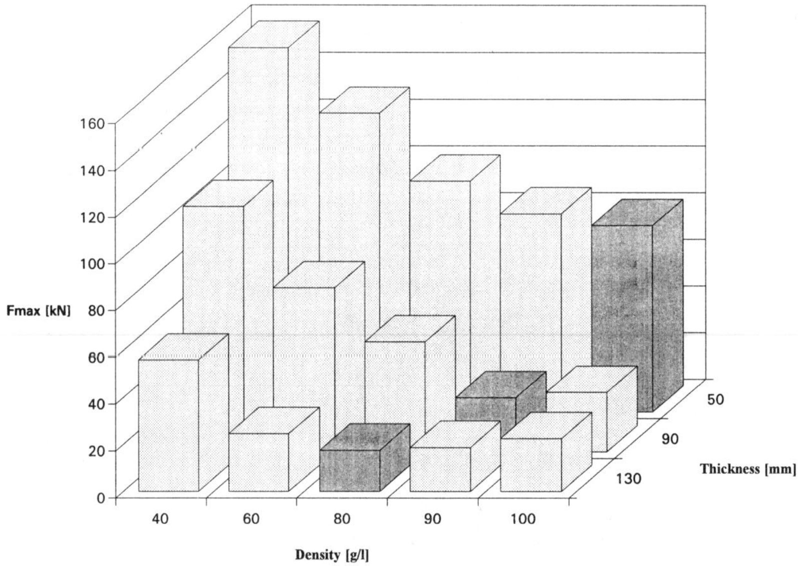

Figure 10 Maximum forces in drop tests using thorax anatomical blocks on to PUR as a function of foam density and thicknes at 1150 J energy absorption

Figures 10 and 11 show the maximum forces measured in drop tests for PUR and PP in various material thickness and density combination. In both figures, the kinetic energy applied is 1.15 kJ. The pair of figures with the best maximum force readings for a given foam thickness is marked in dark colour in each case. To the left of the minimum, the foam components reach solid compaction, while to the right of it the foam hardness is so high that the foam thickness is not optimally exploited.

When the two materials are compared, it is apparent that approximately twice the foam density is required, given an optimum design, for PUR as for PP; for example, at a foam thickness of 130 mm, a density of 40 g/l is required in the case of PP, and a density of 80 g/l in the case of PUR. This weight advantage is not cancelled out by PUR's approx. 10 % better efficiency as compared to PP. In addition, this 10 % is further reduced if efficiency is related not to maximum deformation but to foam thickness (Equation 10). This efficiency can be obtained from Figures 10 and 11 using Equations (6) and (10); a figure of only 0.50 is obtained for 80 g/l, 130 mm PUR, for instance, while efficiency in accordance with Equation (7) is 0.75.

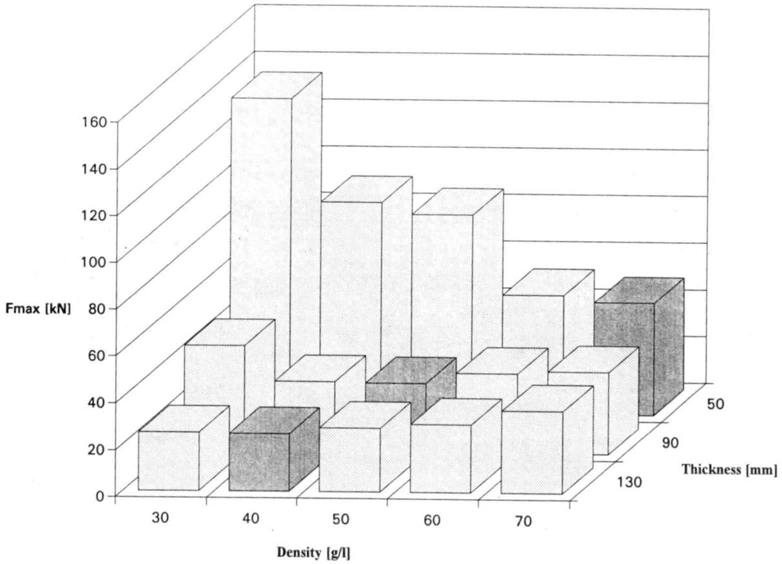

Figure 11 Maximum forces in drop tests using thorax anatomical blocks on to PP as a function of foam density and thicknes at 1150 J energy absorption

In Figure 12, the measurements are evaluated using kinetic energies of 575 J in each case. Due to the lower energy applied, the minima for maximum force occur at smaller foam densities and forces than in Figure 10.

Where even the smallest foam densities of a material are not adequate to achieve the required low level of force for a specific area of the body, such as the thorax, for instance, it will be necessary to utilize cut-outs in the foam which lower the deformation forces proportionately to their dimensions in this area.

An increase in efficiency can be accomplished by means of a rigid covering layer for the foam, distributing the forces to a larger surface area of the foam and thus lessening the geometrical effect at low deformation depths. Figure 13 illustrates this configuration by means of comparison of two force-deflection characteristic plots using the same foam densities but in one case without and in the other with a covering layer. After a steep rise in force to nearly 20 kN, the covering layer of wood fibre composite fractures and the force drops between approx. 10 and 20 mm deformation to less than 15 kN, and then approximates the force plot which results without a covering layer. The kinetic energy is already absorbed at 12 mm less deformation path. This strategy is therefore particularly important where there is little space available and only slight deformation depths are therefore possible.

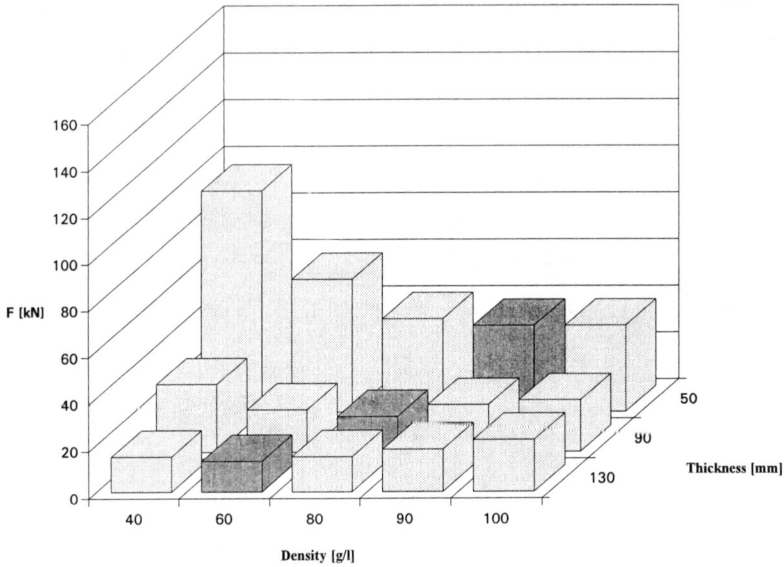

Figure 12 Maximum forces in drop tests using thorax anatomical blocks on to PUR as a function of foam density and thickness at 575 J energy absorption

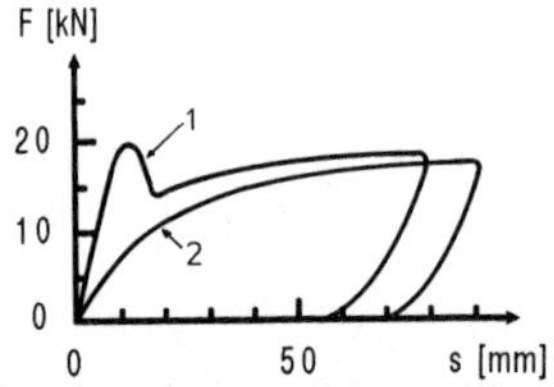

Figure 13 Force/deflection characteristics in drop tests using thorax anatomical blocks for identical PUR foam densities but with (1) and (2) without wood fibre laminate cover (thickness 130 mmm, 80 g/l)

Another method of creating adequate impact absorption zones for the occupant in side-on collisions is to match the deformation behaviour of the internals installed to the dummy. A bass-reflex loudspeaker, the resonance volume of which, taking the form of a blown hollow body, is located directly in the dummy's impact area, is currently being developed for in-door installation at Volkswagen, for instance.

Further applications for plastics in motor-vehicles for energy absorption in case of impact with the vehicle occupants may be anticipated in the new edition of US Standard 201. This law is intended to improve the occupant's protection against head injuries in case of accident, signifying that greatly increased use of plastics and plastic foams may be anticipated in automotive engineering in the future.

Simultaneous Engineering in the development of the new Golf Convertible

Reinhard Pälmer

The previous Golf Convertible, which was based on the Golf I, having achieved production figures at Karmann in Osnabrück of around 388,000 units up to the termination of production in April, 1993, the successor, based on the Golf III saloon, now looks set to repeat this success. The development contract for the version currently on the market was awarded by Volkswagen to Karmann in June, 1990.

The thoroughly founded knowledge and experience which Karmann possesses as a long-term "all-round" system cooperant in the field of development of convertibles and a simultaneous engineering and team-oriented working style made it possible to achieve maximum benefit for the overall concept as early as the planning and implementation stages of the development contract.

Figure 1. Specific plastic parts of the Golf Cabriolet

With the support of ultra-modern dynamic calculation and test methods implemented both at Volkswagen and within Karmann, it was possible to achieve great progress in predevelopment at a very early stage, again as a result of a joint plan for procedure in the development of the bodyshell and plastic components. Great emphasis was also devoted from the very inception in the development of the convertible to recycling-oriented and dismantling-oriented design of the new components (see Fig. 1), a scope which *Skowronnek* [1] already showed in his report on modular units and new applications for plastics in the Golf III saloon.

In view of the fact that the amount of plastics used in the vehicle as a whole is expected to increase further in the next few years, it was particularly important to devote special attention to the needs of materials recycling. This requirement was the reason, inter alia, for the abandonment of several concepts which aimed at enclosing the entire roll bar in foam, as in the predecessor model, since there is no cost-effective method for dismantling and recycling of such a design.

Folding-roof

A folding-roof fixing system modified for visual, installation and servicing reasons was given particular emphasis as a central objective in development. In addition, operation of the folding-roof and fixing of the folding-roof cover was to be improved entirely.

This requirement can now be regarded as achieved. For an outside observer, it is with certainty the visual effect which is the most perceptible. This was achieved by means of a low-level folding-roof stowing compartment which permits the open folding-roof to disappear almost completely into the bodywork (see Figure 2).

Figure 2. Open folding top

This fixing (see Figure 3) makes it possible, thanks to a special folding technique, to use a solid heatable rear window. This design, which is not exactly standard in convertibles, in conjunction with the special folding-roof structure, permits the owner unrestricted use of the vehicle in winter.

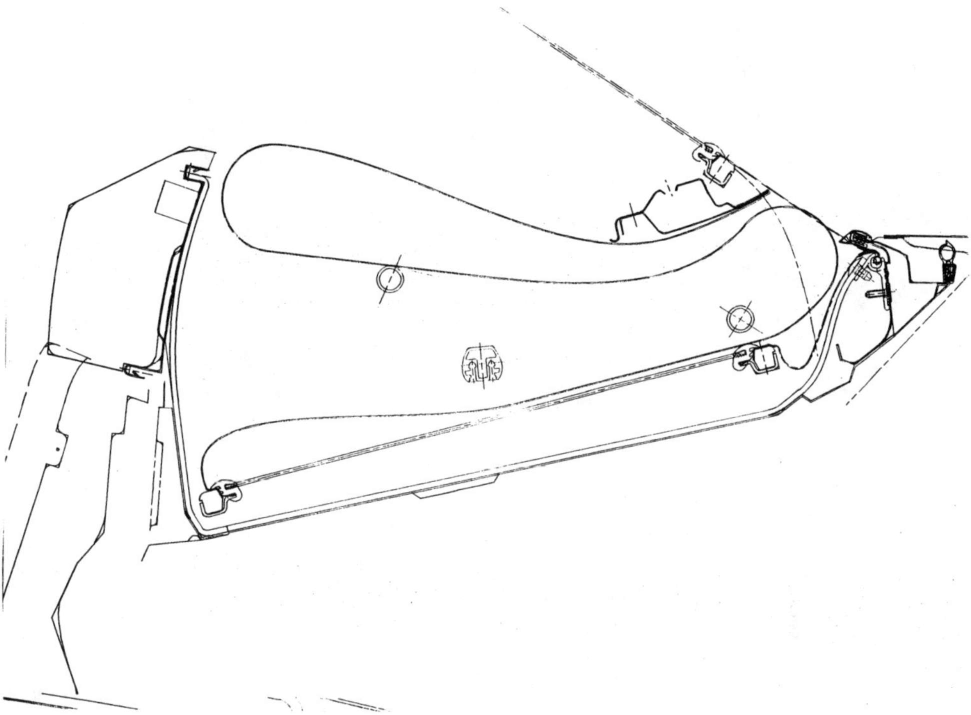

Figure 3. Principle cross section folding-top fitting

In addition, this fixing method makes it possible to fix the folding-roof cover to the exterior of the bodyshell over the open folding-roof with practically no press studs on the exterior of the bodyshell (see Figure 4). This must certainly be regarded as an important step in the direction of user-friendliness.

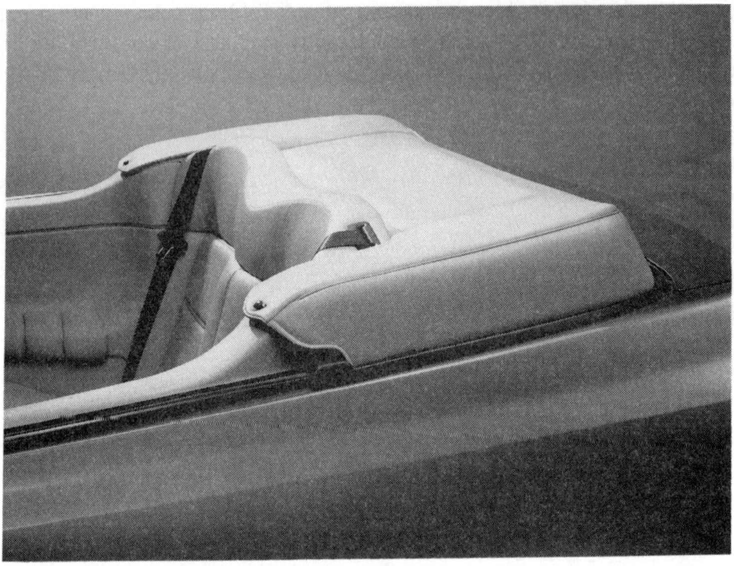

Figure 4. Car with assembled folding-top cover

Sealing elements

The development of the sealing elements was conducted using the same methods and procedures illustrated in 1992 by *Licher* [2], using the example of another convertible model.

The advantages which can be achieved precisely in the development of sealing elements by using the "rapid prototyping" technique will be discussed here using the example of the roof framework sealing element for the Golf convertible. A list of requirements was first drafted, specifying essentially the following activities:

- Development of <u>one</u> element cross-section for <u>all</u> areas (A pillar contour and upper edge of side windows)
- Integration of a wind-deflector lip into the new element cross-section
- Covering of sharp-edged components in the folding-roof framework
- No need for additional fixing elements

Element profile "A" (see Figure 5) was designed on the basis of these requirements.

It became apparent during production of the first specimens that it would be possible to extrude the required element cross-section only by means of the inclusion of additional support ribs within the tubular section.

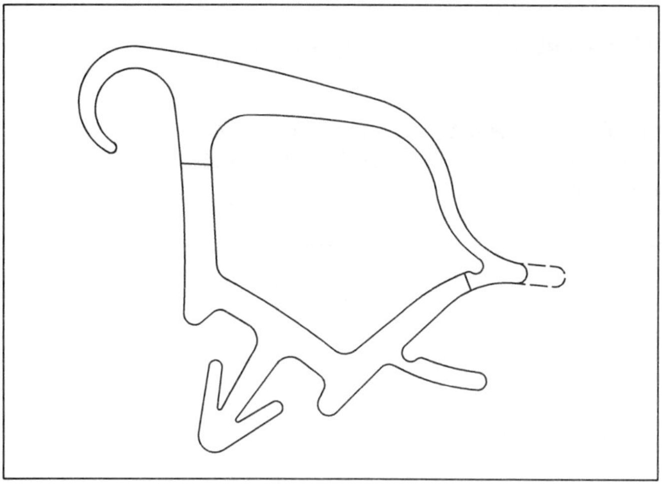

Figure 5. Cross section profile "A"

The revised design resulted in element profile "B" (see Figure 6).

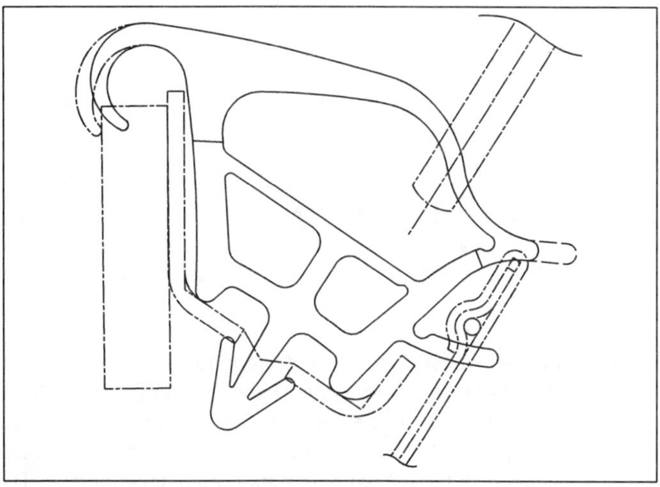

Figure 6. Cross section profile "B"

The following deficiencies were then disclosed when specimens of this variant were installed in the prototype vehicles:

- Attachment feet too weak
- Poor centering of the seal in the mounting
- Difficult and unreliable installation

- Leaks between the element profile and the roof framework
- Internal ribs disrupt moulding-on of the moulded end pieces

In order to eliminate these problems, the attachment feet were revised and reinforced. The internal ribs were redesigned in such a way that they can be removed after extrusion. These modifications resulted in element profile "C" (Figure 7).

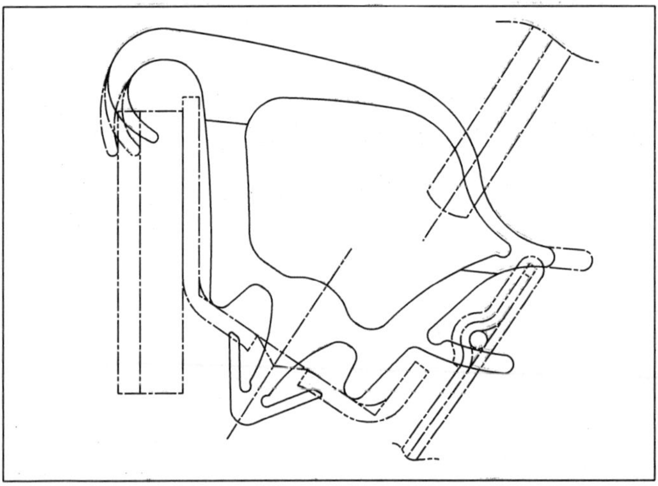

Figure 7. Cross section profile "C"

When specimens of this variant were ready, a higher level of ondulation was observed in the area of the bend at the top of the A pillar. In addition, it became conclusively apparent that this element profile geometry would not be capable of improving assembly and installation circumstances.

The original conception was then completely redefined. It was decided to divide the original element profile. The tubular cross-section became on the whole flatter, a modification which was intended to counter the problem of ondulation. The element profile base was thickened, to permit reliable fixing of the element profile in the newly added aluminium mounting rail. The outer mounting flange "A" was extended and equipped with a support element profile "B" with an integrated wind-deflector lip to support the window. A conventional screw-type mounting with a polymer spreader element "E" in the roof framework was used for fixing of the aluminium rail.

This newly designed element profile incorporated all the features of the subsequent series-production version.

The overall structure of this variant can be seen in element profile "D" (see Figure 8).

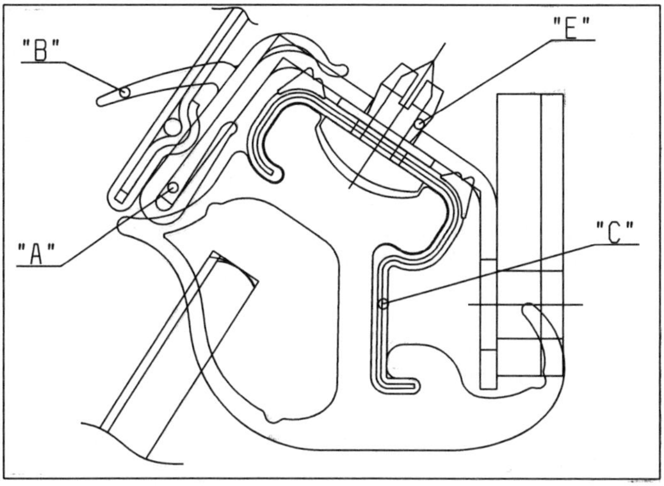

Figure 8. Cross section profile "D"

Thanks to the omission of the inner ribs and the trimming tools for the attachment feet, it could be assumed in the case of this variant that it would involve only a slightly higher financial input.

During subsequent testing of this variant it became apparent that ondulation of the tubular section now no longer occurred. Difficulties did occur with insertion of the tubular section into the mounting rail, but were eliminated by means of slight dimensional corrections.

Water drainage

In the Golf convertible, too, the principle of a special internal water drainage system was used in the form of a system of "intentional" leaks in the rear area of the folding roof with full mechanical car-wash system capability.

A so-called water trap bag, the effectiveness of which had to be assured at all tilt angles in a simulated rainfall system (17° over all sides and the corners of the vehicle) was again used in this system. This component was also made of polypropylene.

Particular importance was attached in this context to simultaneous engineering, since the design of this component required largely empirical procedures.

Immediately upon conclusion of the package phase, stereolithography (STL) components were used to ensure that prototype components of a configuration close to the final series-production version could be made available by the technical departments, with the involvement of the in-house pattern and mould making department.

The rapid incorporation of all the influencing factors also made it apparent that optimization of the procedures, which had resulted in the grouping together of all plastics activities in a single department, had produced significant improvements.

Materials

For materials selection for the trim components, the criterion of single-type material selection, the route pursued in the development of the saloon, was applied, with polypropylene being used for most applications.

It was necessary, here, however, as had already been the case in development of the saloon, to make special allowance for polypropylene's low resistance to scratching in exterior applications. This resulted, for example, in replacement of PP with ASA being necessary in the case of the exterior B pillar trim, the rear trim insert and the folding-roof closure panel.

Since PP can at present be made adequately resistant to scratching only by means of an extremely expensive painting process, there is obviously here a basic challenge to the raw material manufacturers in terms of achieving a significant improvement.

Figure 9 Rear side trim with new upper part and modified gap line

Rear side trim

In the design of the convertible's rear trim (see Figure 9), it was necessary to convert the heights dictated by the retracted folding-roof framework into a visually attractive component suitable for moulding.

It became apparent, entirely in accord with the philosophy of Simultaneous Engineering, that it is increasingly important for both quality and cost reasons to orient work around technical production exigencies even during the design and styling phases.

We should, at this point, draw special attention to the very successful cooperation between Volkswagen's design studio and the Karmann technical departments involved, and no less, to the active cooperation of the sub-suppliers.

The sum of all the activities mentioned above made it possible to significantly shorten development times and enhance development quality. These advances later became clearly visible at the start of series production, in the form, for instance, of a reduction in the times required for reworking and adjustment, and in quality audits.

In total, the procedures described must be regarded as clearly pointing the way for reduction of lead times in projects for the future.

References

1 Skowronnek J.W.: Modular assemblies and new plastics applications in the new VW Golf. Paper given at the Annual Conference Plastics in Automotive Engineering, Mannheim 1993, VDI Verlag, ISBN 3-18-234165-0, p. 11

2 Licher, S.: Applications for plastics using the example of the Renault R 19 convertible hood. Paper given at the Annual Conference Plastics in Automotive Engineering, Mannheim 1993, VDI Verlag, ISBN 3-18-234165-0, p. 83

Plastics applications for the Golf Variant, taking into account new quality targets

Axel Balk

In contrast to the stagnating - and last year actually receding - total European car market, sales in the station wagon sector have recently been increasing significantly.

In the past four years, the rate of increase in station wagon sales has been more than 40 % in Germany and across Europe as a whole.

This trend will almost certainly accelerate with the introduction of the new Golf Variant, the station wagon version of Europe's most successful car, the VW Golf.

Figure 1 Golf Variant

Vehicle conception

The Golf Variant is a vehicle which covers a very broad spectrum of owner requirements. It is, on the one hand, a versatile and attractive station wagon /saloon for family purposes, travelling and leisure, offering a large amount of luggage capacity. On the other hand, it is also a rugged and reliable commercial vehicle with a high level of functional utility.

The principal concentration in sales is unequivocally in private use sector (approx. 90 %). Figure 2 shows the generous and convenient luggage compartment.

Figure 2 The generous, comfortable luggage compartment of the Golf Variant with its carefully matched finish, in the version without a luggage compartment cover

The Variant is based on the Golf saloon and has been completely redesigned from the B pillar to the rear. The vehicle has been lengthened by 320 mm, provides 1425 l of luggage capacity and offers an unobstructed width of 1000 mm between the wheel arches.

Development and implementation of plastic applications particular to the Variant

The new plastic components developed specially for the Variant are discussed below.

Figure 3 The interior trim of the Golf Variant

"Tried and proven" parts and methods provided the starting point: The trim components in the front area of the vehicle are the same components used in the Golf saloon. The standard roof lining and sliding-roof lining did require modification, however.

Interior trim components specific to the Variant were used from the B pillar backwards, i.e., in the rearward direction after the driver and passenger doors, with the exception of the rear door trims.

The material specifications for the Golf saloon and the Golf Variant are identical.
Figure 3 shows a side-view of the interior with the seat folded forward and the back of the rear seat folded down. All thermoplastic trim components in the luggage compartment consist of talcum-reinforced PP-EPDM, with the exception of the lock carrier cover.

Lock carrier cover

For reasons of susceptibility to scratching, this particularly heavily exposed trim component has been produced not in polypropylene but instead in ABS, with embossed runners, see Figure 4. Fixing is accomplished by means of undetachable "fir-tree clips". Sophisticated design of the slots in the bodyshell lock carrier and the "floating" mounting of the clips in the lock carrier cover made it possible to balance out the bodyshell construction tolerances and longitudinal expansion and contraction of the component under exposure to climatic and temperature fluctuations from + 90 °C to - 40 °C.

The lock carrier is shown in figure 5.

Figure 4 Lock carrier cover, Golf Variant

Figure 5 Lock carrier, Golf Variant

The design did, however, initially involve the disadvantage of concealed assembly. In addition, aligned assembly was necessary to achieve the same gap width on the left and the right at the point of transition to the side luggage compartment trim. Following complaints from the final assembly shop, a round dome to permit visible assembly and lateral centering was integrated into the central area of the lock carrier cover. A centering hole was added to the lock carrier on the bodyshell. The lock carrier cover with integrated tubular pin and the lock carrier centering hole are shown in Figure 6. The integrated tubular pin is shown as a detail in Figure 7.

Figure 6 The lock carrier cover of the Golf Variant with a tubular pin for ease of assembly and lateral centering

Figure 7 Detailed view of lock carrier cover with the integrated tubular pin, Golf Variant

A uniform and parallel gap on the left and right is assured, since centering and attachment of the left-hand and right-hand luggage compartment trim also take place at the lock carrier.

Wheel housing cover

Figure 8. The wheel housing cover of the Golf Variant is made of durable PP fleece

Resistance to scratching was an additional reason for not integrating the also heavy duty wheel housing covers (narrowest point of the luggage compartment surface) into the side trim but instead making them from a durable PP fleece, see Figure 8.

Other reasons in favour of a PP fleece wheel housing cover were:
- Maximum luggage compartment width, since the wheel housing cover (fleece) adds only minimal thickness
- Airborne-noise absorption
- Bodywork transmitted noise absorption in the wheel housings, since the PP fleece is attached directly to the sheet-metal wheel housing material.

Due to the accommodation of the McPherson strut in the upper section of the wheel housings, the cavities below the wheel housing cover were dealt with by means of separate packing as the most cost-effective alternative, see Figure 9.

Figure 9. The expanded PP packing beneath the wheel housing trim of the Golf Variant

PU foam-chip aggregate composite components were originally planned for this application. The material was changed to particle-filled PP foam, however, for weight-saving reasons (0.8 kg per vehicle) and because of the requirement for the greatest possible level of type-pure recycling of the trim components.

Pillars and roof framework trim

The trims for the A pillar and B pillar are the same as in the Golf saloon. The C and D pillar trims and the trim strip for the roof lining in the rear section of the car are specific to the station wagon and have their own geometry, see Figure 10.

The fixing system, taking the form of floating fir-tree clips with offset plates and insertion tips were also taken from the saloon, see Figures 11 and 12. Also as in the Golf, the fixing elements are airborne-sound insulated by means of slide-on PU foam mouldings, see Figure 13. The geometrically secured gap-free transition from the pillar trim to the luggage compartment trim was achieved by means of sturdy pointed snap-in hooks in the corners of the pillar trims. Figure 14 shows the hooks used on the D pillar trim.

Figure 10. Pillar and roof frame trims paticular to the Golf Variant

Figure 11 Floating fir-tree clip with offset webs, holow tip and attached PU moulding, Golf Variant

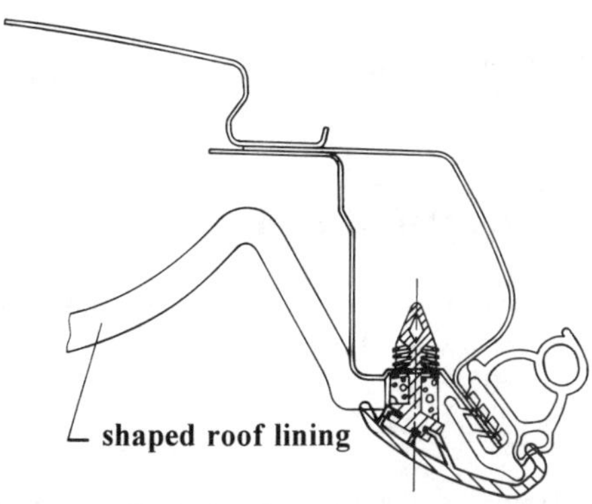

Figure 12 Securing of the roof frame trim in the region of the door seals on the Golf Variant

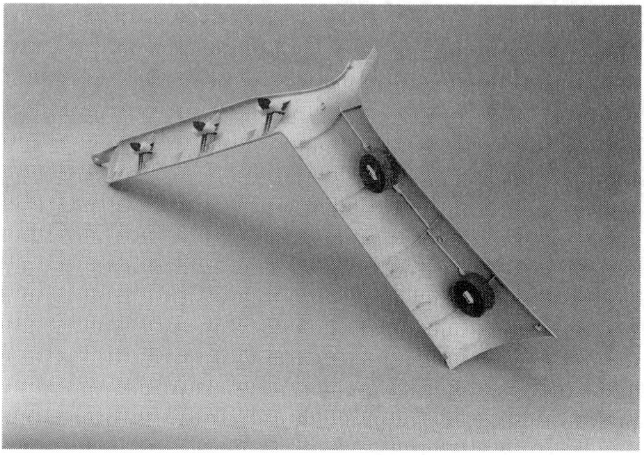

Figure 13 The securing system of the pillar and roof frame trims in the Golf Variant, with attached PU foam mouldings to prevent wind noise in the passenger compartment

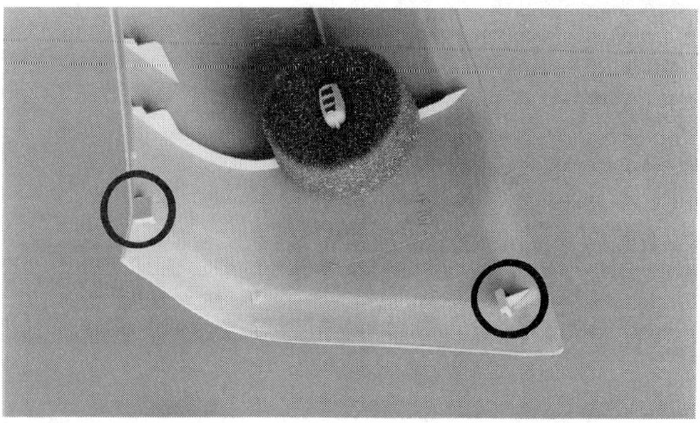

Figure 14. Detailed illustration of the D-pillar trim on the Golf Variant. Sturdy locating hooks for a flush, seam-free fitting to the luggage compartment trim

Left-hand and right-hand rear luggage compartment trim

The rear side luggage compartment trims take the form of easily installable and removable one-piece injection-moulded thermoplastic components with clipped-in ventilation grills. Function-oriented design, with due allowance for the tolerances in the bodyshell, made a single-piece design possible, despite the component's large size.

The geometry of the side trim and the floating design of the fixing system also permits exposure to temperatures from + 90 °C to - 40 °C.

Simple installation is made possible by means of the guidance system in the lower window edge area and the visibility of the remaining fixing points. In the finished vehicle, all fixings are concealed by means of adjacent trim components. Figure 15 shows the luggage compartment trim and its fixing points.

Figure 15 Securing of the luggage compartment side trim to the body, Golf Variant

Figure 16 The Golf Variant body shell, with luggage compartment side trim securing elements located on the bodywork

The side section (bodyshell) with steel brackets for fixing of the luggage compartment trim, is shown in Figure 16.

Figure 17 shows the attachment of the luggage compartment trim around the lower window edge.

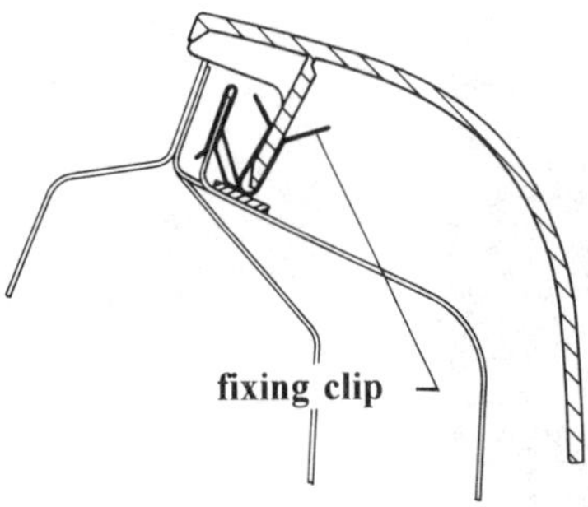

Figure 17 Push-in securing of injection-moulded luggage compartment side trim securing rail, Golf Variant

Fitted stowage box lid

In order to achieve the design target of a fitted stowage box lid and a cost-effective fixing method while nonetheless assuring ease of user-operation of the lids, the hinge axis on the left-hand lid has been unconventionally shifted downward out of the actual lid area, see Figures 18, 19 and 20.

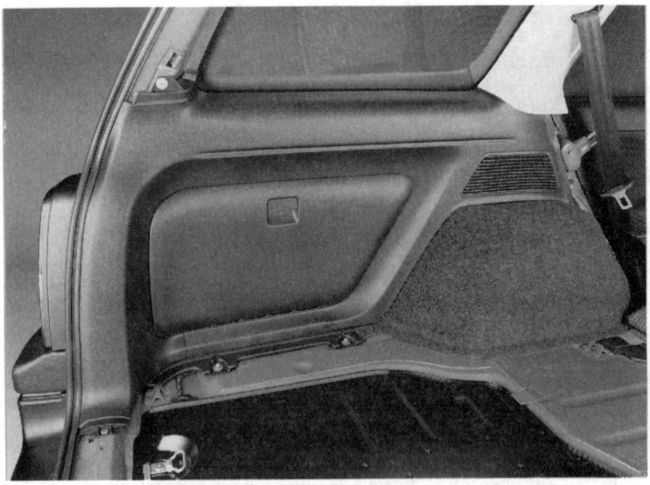

Figure 18 Fitted stowage box lid in the luggage compartment side trim, Golf Variant

In order to eliminate the problem of an irregular gap, the lid halves have been mounted directly onto the luggage compartment trim. Geometrically secured gap-free closure in all temperature ranges had to be assured.

Figure 19 The stowage box lid hinge axis has been lowered.

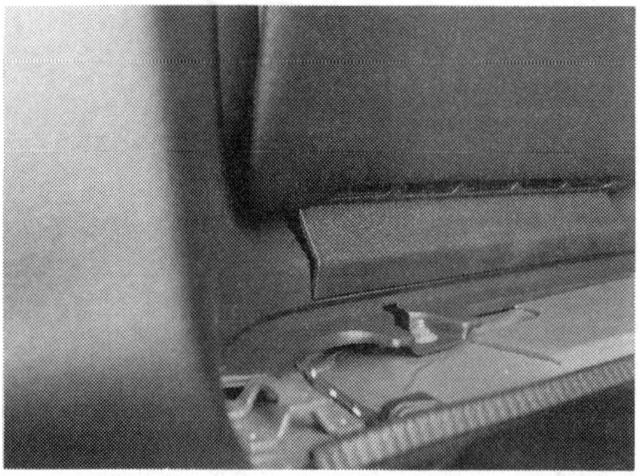

Figure 20 The stowage box lid slightly open

Due to the complex 3D geometry of the moulded components involved, this could not be achieved without further effort. Optimization of the moulding parameters was not sufficient to solve the problem. In addition, it was necessary to match the contours of the lid halves precisely to those of the luggage compartment trim by means of several abrasive operations, achieved by means of tool modifications.

Secure lid closure was achieved by means of a simple system. The robust tongues on the lids engage in clipped-in steel latches in the luggage compartment trim, see Figures 21 and 22. The lids are drawn up against the trim by a special locking tongue design. The necessary closure and opening forces can be changed by adjusting the latch angles.

Figure 21 Stowage box lid securing system; cutaway view showing fibre fleece to prevent wind noise

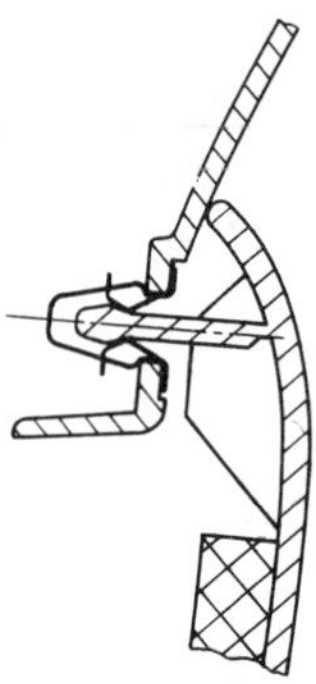

Figure 22 Stowage box lid locking tongue clipped into the luggage compartment side trim steel clamp

The advantage of this latching system can be found in its simplicity and insensitivity to temperature. To permit access to the stowage boxes, the lids can be removed without difficulty in any position and replaced without it being necessary to empty the luggage compartment.

Luggage compartment floor lining

The luggage compartment floor lining textile also consists of PP fleece, see Figure 23.

PP is also used as the material for the welded-on runners and the securing rail for attachment to the floor.

Figure 23 Luggage compartment floor lining on the Golf Variant; fibre fleece, polypropylene runners and securing rail

Tailgate trim and rear bumper

Use of internal gas pressure process

In order to avoid ondulation of the tailgate trim around the lower window edge, the trim has been reinforced in this area by means of a gas duct, see Figures 24 and 25.

Figure 24 Tailgate trim

Figure 25 Tailgate trim (sectional view of channel formed using internal gas pressure process)

For the same reason, the visible upper band of the rear bumper was also reinforced by means of a gas duct.

This hollow cross-section serves to preserve the geometrical stability of the upper edge of the side component, see Figures 26 and 27.

The internal gas pressure process had not up to this time been used in a bumper mould of this size, at least in Europe.

Figure 26 Bumper corner on vehicle

The problem of attachment of the duct is heavily dependent on the geometry of the component. The danger of sink spots in the visible area occurs, but it was possible to eliminate it by means of a slight undercutting and power removal from the mould.

Despite the special demands made on the tightness of the closure in the tool, the use of the gas duct is a good solution for reinforcement and avoiding the need for additional components.

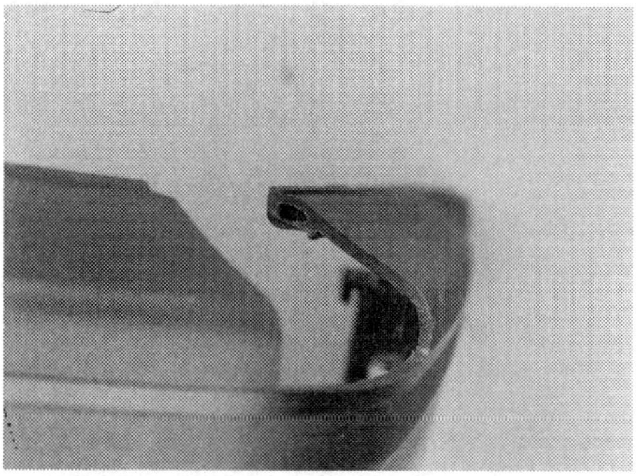

Figure 27 Bumper, channel formed using internal gas pressure process

Achievement of significantly higher quality objectives

The following parameters were selected, in order to achieve Volkswagen's objectives of incorporating into the Golf Variant a greater impression of quality and higher component and manufacturing quality levels:

- Start of zero-series production only once a minimum grade of "3" had been achieved for both VW-produced and bought-in components, i.e., the VDA "Supplier Assessment and Initial Sample Testing" code of practice approved only components produced on series tools using a series-production technique. Divergences were permitted only in the case of colour and grain.

- The condition for scheduled start of series production was an overall vehicle audit grade of not less than 1.8.

The audit grade is a company-internal evaluation system for the delivery quality from the customer's point of view. The grading system is borrowed from the German school grading system, with grade 1 = very good, grade 5 = deficient.

- An approval consisting of members of the works management and the technical development, production and production planning departments was formed. As early as the prototype phase, this body set requirements for further optimization to meet customers' expectations in terms of technology, quality and visual appeal.

- During preparation for series production, production start-up quality was achieved by means of strict pursuance of sub-objectives by the VW board, by means of initially weekly, subsequently daily, coordination meetings involving all departments and areas. Any problematical items were identified along with the persons involved and solved by means of supra-departmental intervention.

The following preconditions were created in order to achieve the high quality and start-up objectives set:

1. 100 % CAD interface definition

The surface of the trim components was completely defined. No prototype models or prototype model replicas were produced for the suppliers; instead, only the data set for the surfaces was provided.

2. Data control model

The component surface data generated on the screen was milled for checking and detection of any surface defects and for design acceptance on a 1:1 scale and then compiled to form a so-called data control model.

Figure 28 Data control model

The data control model was painted to improve the visibility of any surface defects. Figure 28 shows the data control model milled on the basis of surface data for the interior.

Any geometrical changes necessary for technical or aesthetic reasons were implemented directly on screen. Milled model sections were also generated to verify and/or confirm the surface data modified on the screen, see Figure 29.

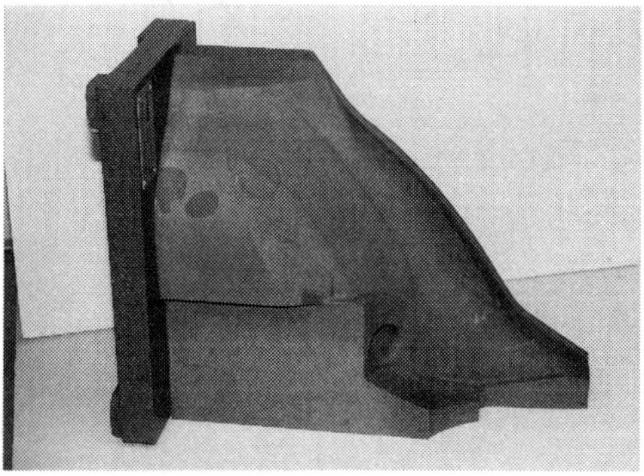

Figure 29 Milled model section

3. Function cubing

The installation of trim components into a nominally dimensioned model with continuous updating with the very latest bodyshell data was the precondition for component approval and for the detection of dimensional divergences.

4. Function dimensioning

Intelligent and consistent dimensioning of the bodyshell and trim components to assure fitting accuracy in the finished vehicle.

5. FPIA

Consistent use of Error Potential Influence Analysis.

6. Fixing of plastic components

Floating design of defined fixing elements making allowance for temperature effects across a + 90 °C to - 40 °C range, and for the manufacturing tolerances for the trim components and the bodyshell itself.

7. Final tool-definition series components

Deadline: Not later than ten months prior to planned start of series production; Series-production components for final tool definition had to be submitted by this deadline.

Summary

Plastic components in motor-vehicles are by no means new. The Golf Variant, too, makes use of tried and proven materials. What is new, however, is the consistent and intelligent pursuit of higher quality standards which not only necessitate the use of modern design methods and the supporting management strategies, but also oblige raw materials suppliers and component manufacturers to pursue continuous improvement in their manufacturing and processing systems in order to achieve precision and high-quality surface finish.

Innovative plastics applications on the Porsche 911 Carrera

Guido Horsch

Development objectives and special features in the development of the new Porsche 911 Carrera

In 1993, the Porsche 911 Carrera celebrated its thirtieth birthday. The latest embodiment of the Porsche 911 Carrera (Figure 1) was presented among the other 1994 models at the 1993 International Motor Show (IAA) in Frankfurt. With 350 000 sales in thirty years, this sports car is the central focus of the Porsche range, constructed at a rate of ten to twenty thousand vehicles per year.

Figure 1 The 1994 Porsche 911 Carrera

The Porsche 911 Carrera has, since its market launch, always been a timeless vehicle with a high potential for further development: In the latest model, too, continuous refinement, always incorporating the most modern materials and processes, ensures harmonization with all future requirements, regulations and legislation.

The design targets for the new Porsche 911 Carrera, such as flowing lines, smooth transitions, integrated front and rear bodyshell assemblies, a fitted rear spoiler, inconspicuous underbody trim and spoilers, were achieved by means, inter alia, of the optimized utilization of polymeric materials. The flowing lines and front and rear bodyshell sections integrated into the overall contour of the vehicle (Figure 2) necessitated maximum surface, paint and stability qualities in the material selected. Above all, the degree of gloss and the paint finish texture had to be matched with the paint-finish quality already achieved on the metal surfaces, due to the need for integration of the front bodyshell section into the overall bodyshell form.

Figure 2 Front bodyshell section integrated into the bodyshell line of the Porsche 911 Carrera

The other targets were defined at the vehicle conception stage:

- Maximum agility with maximum possible active and passive safety
- Sports car road behaviour and best possible long-distance driving characteristics
- Unique vehicle character and
- High longevity.

These were achieved by retaining the air-cooled rear-mounted horizontally opposed engine, reducing the coefficients of air-resistance c_w, $c_{a,front}$ and $c_{a,rear}$ by means of a complete underbody enclosure and the retractable rear spoiler, the consistent use of lightweight engineering materials and by means of the fully zinc-plated steel bodyshell [1].

The principle development emphases were placed on the following items:

- Attractive successor to the 911 Carrera 2/4 entirely in the tradition of the Porsche 911
- Lightweight construction
- Sports car performance with a low fuel consumption
- Cost-optimized design and production potentials
- Exemplary active and passive safety systems
- Maximum possible environmental safety and conservation of resources
- Recyclability and potentials for use of secondary materials.

The environmental safety of the methods and processes used, the disposal potentials necessary for any by-products created, the minimum possible emissions from the materials used and repairability were applied as the criteria for selection of materials.

The following basic requirements applied to the materials selected:

- Assured fulfillment of function throughout the life of the vehicle
- Operating characteristics suitable for all-weather everyday use
- High longevity and component reliability
- Complete process and manufacturing safety
- High quality
- Conformity with the relevant legislation and regulations
- Environmental safety.

These criteria must be assured throughout the production and vehicle-life period; a further requirement was that it should be possible to use an enhanced-performance vehicle for motor-sport purposes without radical modifications to the bodyshell, steering or suspension. This derivative vehicle, with enhanced performance and greatly reduced weight, is to be used in the 1994 Porsche Supercup competition.

It proved possible in the new Porsche 911 Carrera to again increase engine output as compared to the previous model with only a minimal increase in weight, and to significantly improve road-holding by means of an innovative rear suspension and axle design [2]. The high-performance horizontally opposed engine is installed partially encapsulated in the extremely tight engine compartment at the rear of the vehicle. This increases the exposure of the engine components and, above all, of plastic components located in the vicinity of the engine (e.g. the rear lights, rear spoiler and all plastic engine components), due to the high temperatures generated; this exposure is, in toto, significantly greater than in the case of front-engined models.

The high proportion of lightweight materials, such as aluminium, magnesium and plastics, permitted the integration of a broader scope of safety functions and various increases in comfort and convenience while retaining approximately the same weight as the predecessor model.

Particular attention was devoted to lightweight construction of the engine, steering and suspension, in order to reduce the weight of moving parts. All these efforts resulted in the use of various lightweight materials in these different areas: mainly aluminium, in the case of the suspension and steering, both aluminium and plastics in the engine, high-tensile steels and plastics in the bodyshell, and primarily plastics in the interior.

Special features of the Porsche 911 Carrera

The Porsche 911 Carrera is an all-weather sports car for everyday use which, compared to mass-produced vehicles, is manufactured on a small series-run basis and is prepared for various model versions. In addition, allowance must be made for power increases to permit the car's use in motor-sport. The low total production numbers and the number of individual variants necessitate maximum possible flexibility and potential for further development of the individual components. In general, plastic components are adopted without alteration, since modified or alternative tools would result in improportionately high component costs. Where such components have to be modified, designs which can be achieved with simple tools are given preference.

The materials used differ only in their places of use, and not in type, from those used in mass-production vehicles; above all, aluminium and plastics are used as a substitute for steel.

The plastics used are therefore comparable to those used in mass-production vehicles, in the case, for instance, of:

- PE fuel tank
- PP+EPDM front and rear bodyshell sections and front underbody sections
- PP-GMT underbody enclosures
- PP wheel housing shells
- PP fluid tanks
- PP and PA engine compartment components.

All plastic components weighing more than 100 g are marked with their material code in accordance with VDA Recommendation 260, in order to facilitate material identification.

Plastics applications in the bodyshell, steering/suspension, engine and electrical systems

To achieve the lightweight construction of the vehicle as a whole, it is important to use the best possible material for each individual component, in order to obtain the minimum necessary weight. Plastics, the adaptability of which to the diverse requirements of vehicle engineering produce significant advantages over other materials, are used for many components. In the Porsche 911 Carrera, it is above all the potentials and advantages of freedom of geometry, lightweight construction and cost-effective manufacturing which are decisive for the many and varied applications for high-quality polymers. In addition, the integration of a number of functions into a single component which is also possible permits a reduction in the number of components required.

These advantages of plastics over other materials resulted in a high level of utilization of polymeric materials in the Porsche 911 Carrera's predecessors.

The weight-saving materials used in the Porsche 911 Carrera are, above all, light metal alloys and plastics. The vehicle's steel and iron contents together now make up less than 50 %, while its aluminium content has risen to 17 %, and its plastics content to 15.6 %. This signifies the use of a total of 213 kg of plastics in the new Porsche 911 Carrera (Figure 3).

The plastic components are divided as follows between the individual sectors of the vehicle:

- 154 kg (72.2 %) of all plastics in the bodyshell (interior and exterior)

- 35.3 kg (16.6 %) of all plastics in the electrical system
- 18.2 kg (8.6 %) of all plastics in and around the engine and
- 5.5 kg (2.6 %) of all plastics in the suspension and steering systems.

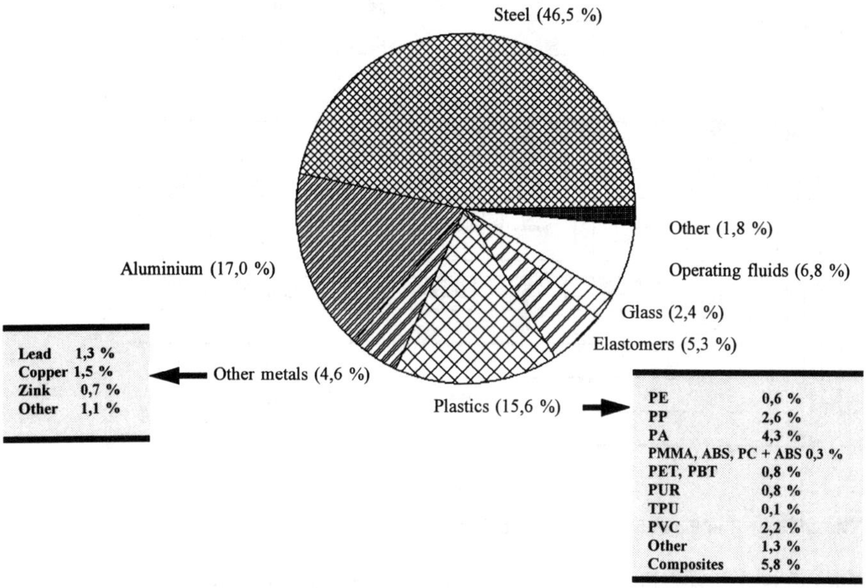

Figure 3 Materials used in the new Porsche 911 Carrera

The most important uses in weight terms are for insulating elements, underbody protection and underseal, and interior and exterior applications in the bodyshell. At the start of series production, various components and masks made of recyclate plastics are used, such as the headlamp guides and the luggage supports on the back of the rear seat; other such components are currently being tested.

In comparison to its predecessors, the Porsche 911 SC and the Porsche 911 Carrera 2/4, the decrease in the car's steel content and the continuous rise in its aluminium and plastic content is conspicuous, see Figure 4. This can be explained both in terms of the replacement of steel components by aluminium and plastics, for instance in the case of suspension/steering components and the fuel tank, and in terms of the use of new comfort and convenience-enhancing components.

In addition, the material for certain components which already use plastics has been changed to a different plastic, as in the case of the rear lights and airbags, for instance. The rear spoilers are now entirely in glass-fibre-reinforced (internal lattice) and mineral-reinforced (framework) polyamide, replacing a frame consisting of a PC+ABS blend with an adhesively bonded glass-fibre-reinforced polyamide lattice, thus achieving the use of a single plastics family, and it has proven possible to convert the front and rear body components to mineral-reinforced PP+EPDM, to replace glass-fibre-reinforced and mineral-reinforced TPE+PBT [3].

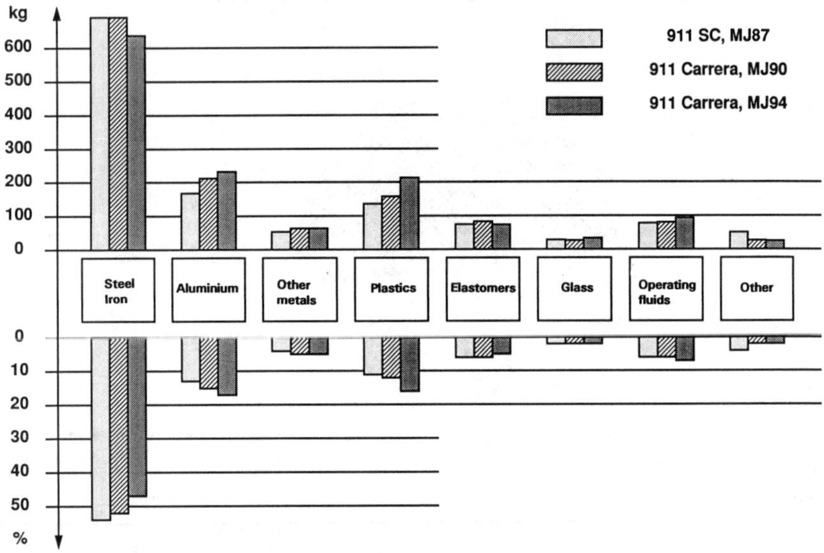

Figure 4 Materials used in the Porsche 911 Carrera in various years

In addition to the cost advantages, the reasons for these changes included:

- Recycling requirements (rear spoilers and airbags)
- Lightweight construction (front and rear bodyshell sections, airbags)
- Process improvements (rear lights)

Above all, the conversion from painted TPE+PBT to painted PP+EPDM for the front and rear body sections made possible weight savings of approx. 20 % for these components, as well as significantly reduced component costs.

Requirements made on plastic components and materials selection in the individual sectors of the vehicle

The principal applications for plastics in the bodyshell are on the exterior (Figure 5).

- Painted front and rear body sections, front underbody components
- Underbody enclosures and sill trims
- Painted rear spoiler
- Painted exterior mirrors
- Wheel housing shells
- Fuel and activated carbon tanks
- Heating and air-conditioning system housings.

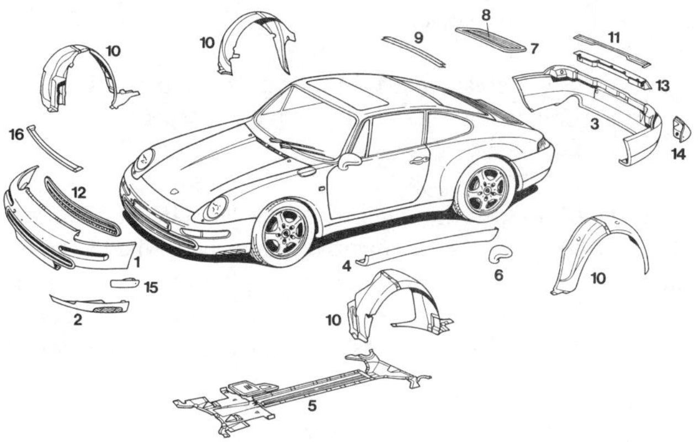

No.	Name
1	Front body section
2	Lower front
3	Rear body
4	Sill trim
5	Unterbody enclosure
6	Rearview mirror housing
7	Rear spoiler
8	Rear spoiler grill
9	Roof strip
10	Wheel housing shell
11	Roof mask strip
12	Ventilation grill
13	Rear mask
14	Rear winker housing
15	Front winker housing
16	Weatherstrip

Figure 5 Some of the applications for plastics in the bodyshell exterior

The front and rear body components, the front underbody parts, the sill trims, the individual components of the rear spoiler and the wheel housing shells are produced by means of injection-moulding, while the underbody enclosures consist of GMT, the tanks are produced by means of blow-moulding and the exterior mirrors using the internal gas pressure process.

In the case of the exterior mirrors, the internal gas pressure process made it possible to produce an attractive painted visible component with no mould parting lines (Figure 6).

Figure 6 Exterior mirror produced using the GIP method

In the future, painted plastic exterior skin components will also be conceivable in high-end vehicles. The first such components will be wings and tailgates, once painting can be cost-effectively accomplished to the necessary quality standards.

In all vehicles, the vehicle interior is the principal field of application for plastics. In the interior, plastics can be used in trim components, masks and covers, carpeting, textiles, padding and also in various structural components.

A large number of plastic components are used in the interior of the Porsche 911 Carrera; compared to the previous model, various changes in materials have been made, aimed primarily at reducing fogging emissions. The reduction in the range of plastics types used and the utilization of single-type designs for assembled components also improves the recyclability of interior components which are not directly visible. The most important interior applications are as follows:

- Interior and door trim components, carpeting and mats, roof lining
- Insulating elements
- Heating and air-conditioning system housings, including integrated interior air filters and air ducts
- Seats, seat-belts, airbags and other textile elements
- Dashboard covering, masks, glove compartment
- Knee guards and center console
- Airbag cover.

The materials for the airbags and the knee guard mountings are among those changed in this context.

Figure 7 The interior of the Porsche 911 Carrera

The number and size of the components used add up to a weight of 154 kg. The materials used are almost exclusively thermoplastics, technical components consisting largely of single-type designs, although the visible trim components in the interior in some cases consist of compound elements with high-quality surfaces.

Visual enhancement of plastic trim elements in the interior (Figure 7) is accomplished by means of leather, semi-leather, textile, woven and artificial leather covers. An expanded range of requirements applies to the selection of materials for components to be covered. This can, in individual cases, result in the selection of a higher quality material.

Figure 8 The front axle of the Porsche 911 Carrera

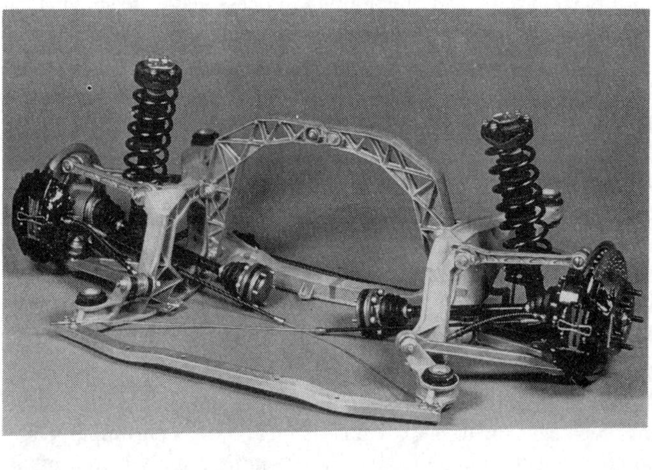

Figure 9 The rear axle of the Porsche 911 Carrera

In the steering and suspension systems (Figures 8 and 9) only minor non-safety-relevant components are made of plastic, since the load-bearing components are subject, as highly safety-relevant items, to extremely high function-fulfillment requirements. Various small plastic covers and pipes are used in this area, however, together with the air-spoilers for the brake disks.

The most important uses of plastics here are:

- Air-guide spoilers for cooling air to the front brakes
- Brake-fluid header tank
- A-arm cover
- Wheel-hub covers

The air-spoilers have been converted from non-reinforced polyurethane to EPDM-modified PP. This material fulfils all requirements, including above all resistance to stone chipping at low temperatures, and also achieves a significant cost reduction.

In the electrical system, plastics are used primarily as covers and housings for electrical components. Major plastics-containing components in this context are, for instance:

- the tail lamp and flasher housings, parts of the headlamp housings
- ABS, control unit and instrument housings
- battery casings and cable sheathings
- plugs, connectors and mountings.

Due to their proximity to the rear-mounted engine, the rear lamp units above all, must be able to withstand elevated exposure to temperature. In the previous model, these requirements were met by the selection of temperature-resistant PMMA and PC types. Problems with susceptibility to stress cracking at the weld between the lens and the housing were the reason for the search for alternative materials.

In the case of the tail lamps, the material and the joining process have been changed (Figure 10). The size of these components and the sharp curvatures of the connecting surfaces necessitate a flexible but nonetheless tight joint between the inner and outer shells. Experience gained with the previous model using mirror-welding under comparable geometrical conditions resulted in the decision to make the housing of talcum-reinforced PP and join it to the highly temperature-resistant PMMA lenses (Type HW 55) by means of a new adhesive bonding system. The high thermal loads are lessened by the use of aluminium heat-protection linings behind the rear section of the bodywork. This flexible bonding system eliminates stress cracking and ensures improved tightness.

Further efforts to achieve lightweight construction are implemented in the engine. In view of the target weight distribution of 40:60, it is necessary to achieve the minimum weight of engine aft of the rear axle.

The diverse uses of plastics (Figure 11) made it possible to achieve not only a significant reduction in weight but also advantages in terms of noise suppression, consumption and exhaust output, component number reduction and the aim of reducing assembly labour input, by means of a modular structure.

Figure 10 Tail lamps consisting of an adhesively bonded PP-TV housing and PMMA lens

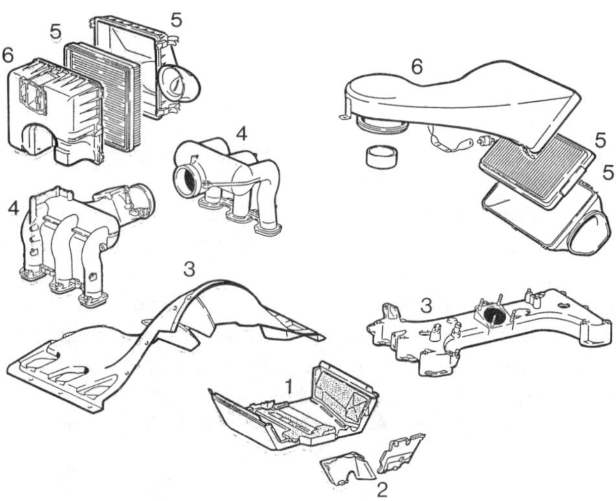

No.	Name
1	Engine guard
2	Noise suppression guard
3	Air intake
4	Intake manifold
5	Air filter
6	Cowl for air filter

Figure 11 Certain uses of plastics in the engine compartment

The most important components are:

- the underbody enclosures and covers
- the intake system (manifolds on the left and right and spacer) and the air-filter housing
- the air flowmeter

- fan housings and air ducts
- the dynamo cover
- the fan impeller and surround for the oil-cooler
- the vacuum accumulator and
- the valve cover.

Adaptation of the engine design made it possible to retain certain components unchanged from the previous engine; in other cases, component geometry was modified. The plastic valve covers are used for the first time. Certain uses of plastics in the engine will be discussed in more detail later.

Environmental requirements for use of plastics

In the new Porsche 911 Carrera, Porsche fulfils all the legislation and regulations applicable to materials throughout the world. In addition, the long production periods mean that extensive preparations must be made for all foreseeable increases in stringency and planned new regulations. Particular attention must be devoted to the following regulations and internal codes of practice, inter alia, in the case of plastics:

- No cadmium content
- CFC-free (and CFC-free manufacturing processes, too)
- Low-fuel-diffusion materials for fuel-contacted components
- Avoidance of solvents for initial and repair painting
- Recyclability
- Component marking in accordance with VDA 260
- Low-emission materials in the interior of the vehicle
- Preferential selection of materials also used by mass-production manufacturers.

Attention is also directed to making as few modifications for specific export destinations as possible. Since Porsche does not produce any of its plastic components itself, close cooperation with component suppliers is necessary in the field of adherence to and implementation of environmental requirements. Porsche is preparing for plastics recycling at two levels. Involvement in PRAVDA (abbreviation for the German title "Projektgruppe Altautoverwertung der deutschen Automobilindustrie = German Motor-vehicle Industry Life-expired Vehicle Recycling Project Group) and harmonization of plastic materials selection with the high-volume manufacturers are the guarantees of optimum materials and logistical recyclability of the materials used.

Innovative developments in the engine compartment

Like the engine of its predecessor, the Porsche 911 Carrera's six-cylinder horizontally opposed engine is air-cooled and installed in the rear of the vehicle (Figure 12). This configuration constitutes one stylistic element of the 911 philosophy. The high specific power output, the demand for ease of maintenance and reduction of component numbers make maximum demands on the plastics used, due to the limited space available and the temperatures which occur [4,5].

Figure 12 The six-cylinder horizontally opposed engine of the Porsche 911 Carrera

The overall plastics content of the Porsche 911 Carrera engine makes up practically 8 % of the engine's weight (Figure 13), i.e., well over twice the 3 % average [6] of most engines. Due to the spatial separation, incidentally, the heating and air-conditioning system is considered part of the bodywork peripherals rather than the engine.

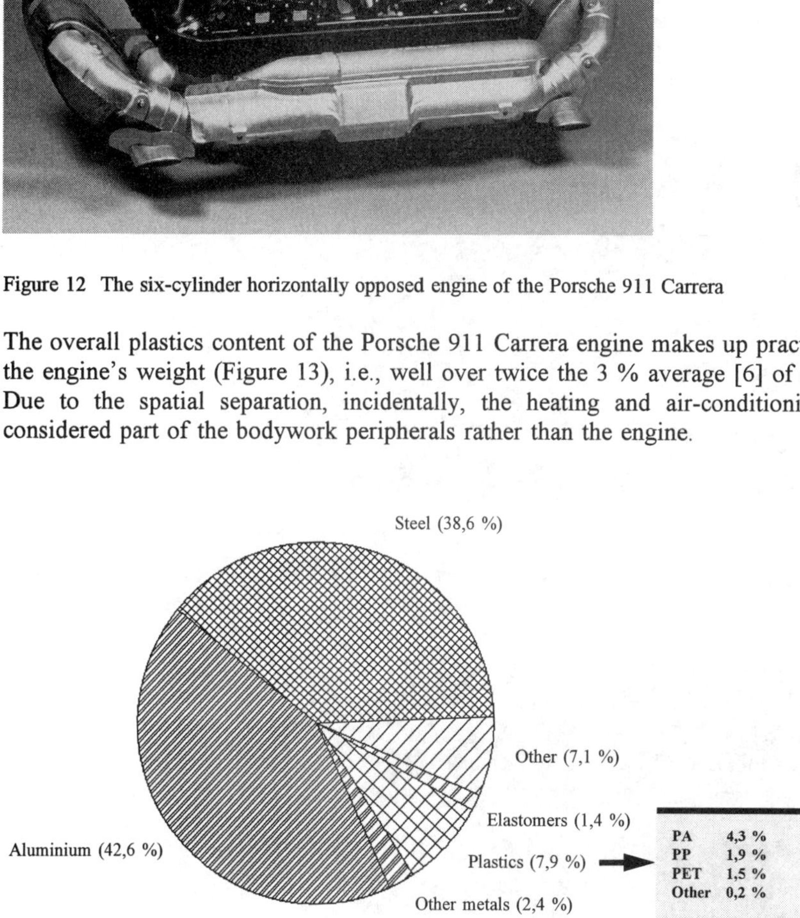

Figure 13 Materials used in the six-cylinder 3.6 liter engine

The variety and size of the plastic components used and the efforts made to achieve lightweight construction in all other components, above all by means of the high-level of aluminium utilization, are the reasons for this. A further reduction in engine weight by means of increased and improved use of plastics via the development of further plastic components is possible and is a current development target.

The advantages and development history of certain uses of plastics in the engine of the Porsche 911 Carrera are described below using various examples.

Valve ("rocker-box") cover

In the new Porsche 911 Carrera, valve covers in glass-fibre-reinforced polyamide, a new plastic component development, are used for the first time (see Figures 14 and 15). These seal

Figure 14 Top and bottom valve cover, exterior

Figure 15 Top and bottom valve cover, interior

the oil-lubricated valve mechanism drive system off from the engine compartment. They achieve weight savings of approx. 20 % compared to the die-cast magnesium covers and approx. 50 % compared to the die-cast aluminium covers used in the previous models, accompanied by significant component cost reductions.

The main materials requirements for the use of plastics in the valve covers are as follows:

- Long-term oil-tightness throughout the temperature range up to 140 °C
- Oil-resistance, both for oil splash (at top) and stationary oil (at bottom)
- Resistance to ageing
- Resistance to corrosion.

In the previous engine, magnesium's inadequate corrosion properties had resulted in a return to aluminium. It was necessary to accept the disadvantage of extra weight, since it was not possible to eliminate magnesium contact corrosion. The only choice for the valve covers for the new engine was therefore between aluminium and plastics.

Initial tests were performed using the test tool on two different polyamides. The suitability of a heavily glass-fibre-reinforced polyamide was assessed against that of a glass-fibre and mineral-reinforced polyamide. Even at this stage, the suitability in principle of the polyamide material for use in valve covers was apparent. The material ultimately selected was a PA66-GF35, the properties of which are approximately mid-way between the two types examined.

In various optimization stages, the rubber-gasket-sealed valve covers were stiffened by means of ribs and the bolting system was improved in such a way that leaks resulting from shrinkage and creep of the material and distortion of the component during manufacture were eliminated. The reduction of manufacturing distortion by means of optimization of reinforcements and of the manufacturing process was the decisive factor in the selection of plastic valve covers. Tightness against stationary oil in the lower areas of the valve covers is assured by means of comprehensive rib-reinforcement and a shorter bolt spacing than in the upper areas. The upper valve cover requires an adequate rib-reinforcement system to achieve tightness against oil splashes, and two additional bolts as compared to the aluminium version used in the previous engine.

In manufacturing the valve covers, it is necessary to ensure the best possible achievable flatness for the sealing frame on the inner side of the covers in order to ensure that the bolt forces necessary for complete tightness do not cause impermissible stresses. The metal inserts used reduce bolt-head surface pressures and the short cycle times achievable cut component costs. The use of the PA66-GF35 material, in combination with optimization of the bolting system, above all in the components corners, made it possible to achieve tightness up to a pressure of 2 bar. Optimization of the bolting system resulted in a total of four new bolting points vis-à-vis the die-cast aluminium valve covers used on the previous engine. The increased installation-work input is balanced out by the significantly lower component costs.

The suitability of the glass-fibre-reinforced polyamide selected has been verified in a large number of tests and experiments; the sealing system and bolting have been optimized, for instance, by means of a large range of FE calculations (Figure 16) [7].

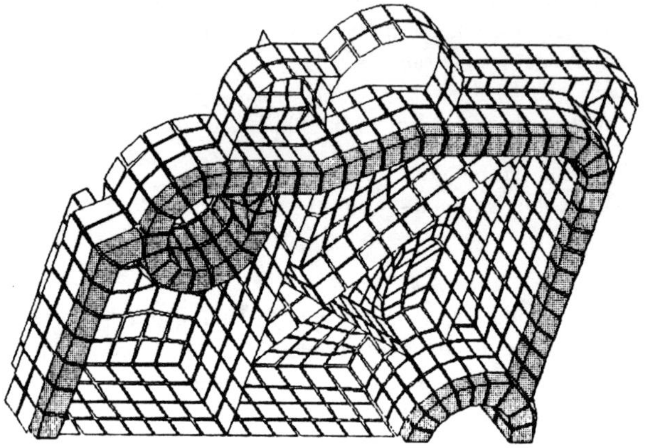

Figure 16 FE grid at the critical angles of the valve cover [7]

Examination in accordance with *Dangelat*[1] of the energy balance of the various available material and design solutions indicated a clear advantage for the total energy balance of the polyamide solution when compared to the die-cast aluminium solution it replaces. This is achieved even on the basis of initial use, without even taking account of material recycling potentials. Both materials can be materials recycled, the total energy balance after the first recycling circuit shifting only slightly toward aluminium.

Intake system

Plastic intake systems functioning on the resonance induction principle with both fixed and variable intake pipe lengths have already been comprehensively investigated by many manufacturers around the world and are used above all in high-end and high-performance vehicles [6,8,9].

Even in the 3.6 l predecessor engine, the weight of the intake system was significantly reduced and component costs decreased by the plastic intake manifold system as compared to the conventional die-cast aluminium solution. Detailed tests were used to establish the suitability of various polyamides as materials for the inlet manifold. The most suitable material proved in the tests to be a PA66-GF35, although the other polyamide types, in some cases with various reinforcements, also met the requirements and were approved as potential alternatives.

[1] Dangelat, M: Ressourcenschonung durch Stoffrecycling unter Voraussetzung von positiver Energiebilanz am Beispiel Zylinderkopfhaube (Saving resources by recycling on the basis of a positive energy balance sheet, taking the example of a cylinder head cover) graduate diploma at Fachbereich Fahrzeugtechnik und Konstruktion TH Zwickau, 1992

The previous engine's intake system was retained for the new engine, since the geometries and configurations involved remained unchanged. The unthrottled intake system (Figure 17) contributes to power enhancement thanks to the lower flow-resistance presented by the smooth internal walls. In addition, intake noise is attenuated by means of a reflection system with an integrated air-filter and intake snorkel. Decoupling of the vibration-endangered attachment components (such as the idling adjuster, control valve and micro-switch) and the manifold as a whole was accomplished by means of rubber grommets which divide these components off from the rigid and vibrating engine.

The main requirements on the material are as follows:

- high fatigue strength
- flange tightness
- ageing and
- thermal stability up to 120 °C

and, in the spacer, additionally

- high geometrical stability, due to the internal resonance flaps.

Figure 17 Complete intake system showing inlet manifold, spacer, air filter and air flowmeter

Manufacture is accomplished using the core-melt-out process. A bismuth core is inductively melted out in an oil-bath after the injection moulding operation has been completed, making it possible to fabricate even undercut cavities.

Guide and tensioning bars

The PA66-GF35 guide and tensioning bars with moulded on polyamide runner linings (Figure 18) for the cam-shaft timing system were also retained from the previous engine. They serve to guide the duplex chains which drive the cam shaft via the layshaft (Figure 19).

Figure 18 Guide and tensioning rails in PA66-GF35

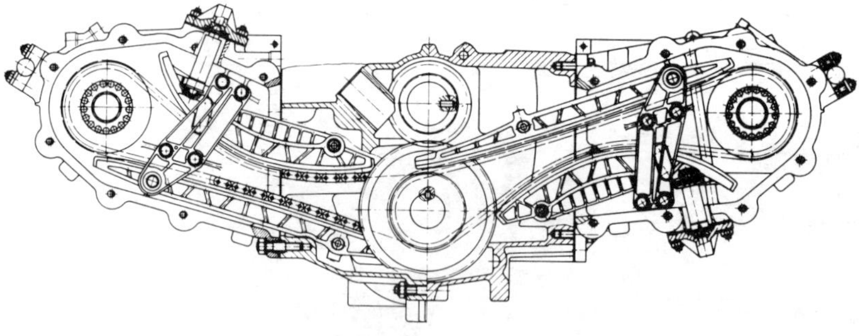

Figure 19 The function of the chain mechanism running over the guide and tensioning rails

In the previous engine, these rails replaced short guide bars and chain-sprocket tensioners. This made it possible, thanks to improved axial chain guidance, to significantly decrease the noise generated by the chain-drive system. In addition, replacement of the short guide bars and chain-sprocket tensioners by the tensioning and guide bars reduced both the weight and component cost of the chain-drive system. It was possible at the same time to reduce the manufacturing and assembly work input by decreasing the number of components and eliminating certain machining steps.

Alternative solutions involving metal rails with plastic runner linings were rejected for wear, weight and cost reasons.

In this application, the polyamide used is subject to the following requirements:

- Abrasion and wear caused by the action of the chains
- Exposure to temperature from oil splash (up to 135 °C) and friction-generated heat.

No significant traces of wear were observed after comprehensive tests. The material's mechanical properties are not impaired by its exposure to load and fatigue strength actually increases after the running-in phase.

Further innovative plastic components in the engine

Two other interesting components in the engine of the Porsche 911 Carrera are the engine underbody enclosure and the top air intake.

The underbody enclosure (Figure 20) serves as a noise-suppression element and also improves the vehicle's aerodynamics by completely enclosing its underbody. Thanks to internal modifications to the engine, the underbody enclosure is no longer required to completely enclose the engine, and replaces several encapsulated components used in the predecessor model. The use of quick-close attachments instead of bolt-type systems also significantly reduces the assembly-work input.

Material requirements include, in particular, resistance to stone chipping, particularly at low temperatures, and thermal stability up to a maximum of 230 °C. In addition, the component must remain geometrically stable even beyond this temperature range and must be achievable at low component costs with the minimum possible weight. These requirements resulted in the development of an underbody enclosure consisting of a glass-fibre-reinforced PET using GMT technology.

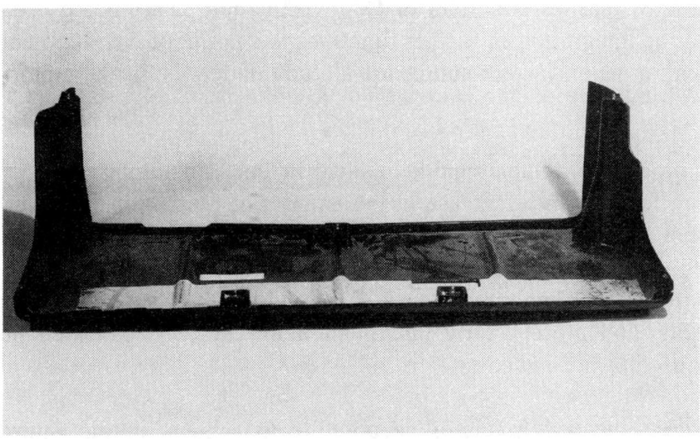

Figure 20 PET-GM30 engine and underbody enclosure

The top air intake (Figure 21) takes the form of a glass-fibre-reinforced polyamide injection moulding. Integrated into it are the rear surface masks; the whole is bolted to the engine. This proximity to the engine necessitates thermal stability up to 120 °C in the material, and a design insusceptible to vibration damage. It proved possible to achieve the plastics version at significantly lower component costs than all the potential sheet-metal and SMC solutions.

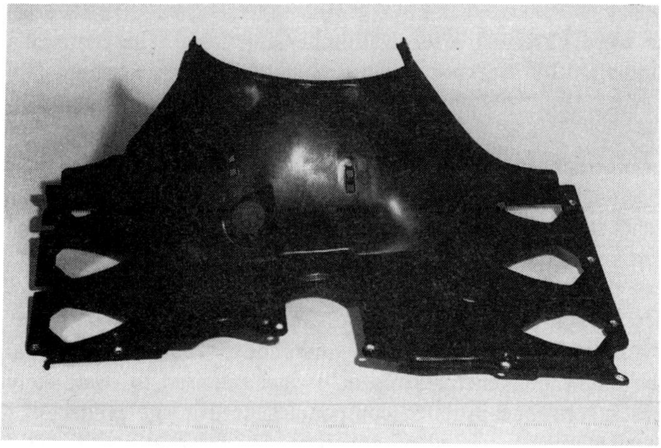

Figure 21 PA66-GF30 top air intake

Development tendencies for plastics applications in the engine compartment

Plastics are certainly capable of making further contributions to weight savings and reduction of the number of components in the engine compartment. Integrated functions in existing components are conceivable, for instance, in the case of the intake system and the water-pump. Intake systems consisting of plastics are either under investigation or are being used by all manufacturers [6,8], and the integration of further functions is a target of development work at present. Plastics bodies for oil and water-pumps are already under test in the motor-vehicle industry [10].

Even more improvements in process and functional reliability in both manufacturing and service will be the precondition for a further increase in the amount of plastics used in and around the engine. Highly temperature-resistant thermoplastics will be preferred to thermosets. Increases in resistance to temperature and creep in the thermoplastics will open up further potential applications to these materials. Components such as oil and water-pump bodies, for instance, thermostat components, and more greatly integrated intake systems can then be developed for series production.

Highly promising plastic components which it will be possible to achieve in the future include:

- oil-pumps with plastics bodies
- oil flow control pistons
- water-pumps with plastics bodies (with an integrated thermostat)
- timing case cover and closures
- coolant, air-conditioning, fuel and oil piping
- thermostat components and
- further refined intake systems.

Intake systems will in future increasingly be equipped with variable resonance systems and longer intake pipes, in order to optimize medium engine-speed performance. It will be possible to integrate additional functions, such as the throttle body, for instance. Even the return routing of hot exhaust gas into the intake pipe is conceivable. In addition, component costs can be further reduced, particularly for short-run production, by means of a built up design consisting of two shell-sections welded together.

The reduction of component costs for complex plastic components by means of new and alternative methods of design and manufacturing processes will also improve the potentials for utilization in lower-cost categories of vehicles.

Fibre-composite components, which have been investigated more intensely in recent years, will not come into greater use in the future, due to the still inadequate level of process reliability, high costs and the lack of large-run series production processes. Individual items for small special production series or high-performance variants may very well be achieved by means of fibre-composite systems, however.

Most development work nowadays is concentrating on extended thermoplastics, the process and functional reliability of which have already been demonstrated on a large number of components.

More highly temperature-resistant thermoplastics will increasingly come into use in the future, provided they can fulfil all requirements regarding resistance to temperature and creep. An overall energy balance can, for instance, then be used to provide additional support in the selection-of-material decision.

Recycling of plastic engine components

In the various investigation projects conducted at Porsche and within the scope of the PRAVDA and FAT activities, certain engine and associated plastic components have been examined for their ease of removal/dismantling and their recyclability [11-13]. Promising results for materials recycling were obtained for the following components, for instance:

- PA intake systems
- PP air filters
- PP coolant tanks.

Binding information on the recyclability of these components can be provided only once sufficient quantities are available from life-expired vehicles and discarded wear parts [11,12].

In general, the following advantages and disadvantages should be anticipated in plastic components for installation on and around the engine:

o Advantages:

+ Single-type designs
+ Expensive but high-quality materials
+ Limited range of variants
+ Repeat utilization possible in reconditioned engines

o Disadvantages:

- Contamination with vehicle fluids
- More complex joining systems
- Functional foreign bodies.

Recyclability of plastic components for the engine and engine compartment sectors is supported by means of corresponding design guidelines and appropriate selection of materials.

Prospects for the use of plastics in short-production-run motor-vehicles

In the development and design of plastic components, principal emphasis is placed, even in the case of low-volume vehicles of 1000 to 10000 units per annum production rates, on maximum exploitation of the material and its lightweight potentials. In materials selection, paramount attention must be devoted to material-oriented component design, functional requirements, environmental safety and component costs. Thanks to their capacity for individual adjustment of their property profile, plastics are the ideal materials for many applications. Once the existing environmental and recycling potential is realized and large-scale recycling processes become reality, it will be possible to recommend polymeric materials without qualification.

Other advantages of plastics, such as:

- potentials for the expansion of the functions of individual components
- enhancement of safety by means of more resilient bodyshell components
- improvement of properties via new production processes
- potentials for optimized lightweight construction

will in future make the use of plastics viable for many other components and expand their potential for further development.

For low-volume production runs, it is development and lead times which are of decisive importance. Simpler tools and the building-up of complicated components from simple individual parts provide good potentials for reducing development and lead times. Even complicated components could then be modified more quickly, easily and cheaply simply by changing individual parts. Energy balances in addition make it possible to draw comparisons between different materials in order to permit selection of the material which offers the greatest energy-economy measured across the vehicle's entire life. In many cases, this material will be a plastic, since the alternatives are often heavier and therefore would lead to the consumption of more additional energy during the vehicle's useful life than they would save during the manufacturing process.

Even more stringent requirements will be additionally made on plastics in the future, including higher resistance to temperature, smaller material thicknesses, systematic manipulation of properties by means of intrinsic reinforcement and new manufacturing processes, for instance.

A further important function in the continued development of plastics is the achievement of environmentally safe production processes and the reduction of pollutant emissions during component manufacture and during the utilization phase, above all in the vehicle interior.

Provided all the basic requirements, such as functional and process reliability, consistent quality and fulfillment of all regulations and legislation can be met, the decisive selection criteria will be component cost and component weight.

A further objective of material selection is the reduction of the need for auxiliary production materials and equipment and the avoidance of escaping additives. The basic requirements for functional and component reliability and safety must be fulfilled during the in-service phase; repair of minor accident damage should be possible.

After expiry of their useful life, it must be possible to either route the plastic components to a commercially available recycling system or to dispose of them without detriment to the environment. A reduction in the variety of types of plastics used and/or the achievement of compatibility of the variants with one another would significantly improve the future situation.

The recycling projects operated by the relevant motor-vehicle and raw materials industries must systematically set up and expand recycling capacity, i.e., economically viable processes must be developed and launched by motivated companies.

There is an increasing need for new plastics applications in low-volume sports vehicles, since there is a requirement for continuous improvement of corrosion-resistance properties and for reduction of the number of components by means of multi-functionality in individual items, despite the increasing demand for lighter vehicles. We may conclude by saying that higher plastics contents in vehicles are perfectly conceivable in both low-volume and special vehicles, provided all the requirements are fulfilled.

References

1 Kahnau,B., Bofinger,G., Melcher, W. and Rohardt, H.: Der neue Porsche 911 Carrera (The new Porsche 911 Carrera), ATZ 95 (1993), October, 1993

2 Kroninger,W., Söffge, F.:Das Fahrwerk des neuen Porsche 911 Carrera (The suspension and steering of the new Porsche 911 Carrera), ATZ 95 (1993), November, 1993

3 Burst, H., Reiche, R., and Wagner, D.: Bumper coverings for the Porsche Carrera 4 - Thermoplast blend as an alternative to PUR-PRUM. Paper given at the Annual Conference on Plastics in Automotive Engineering, Mannheim 1989, VDI-Verlag. pp. 111/133

4 Beer, M., Körkemeier, H. and Zickwolf P.: Der Motor des neuen Porsche 911 Carrera (The engine of the new Porsche 911 Carrera), MTZ 54 (1993), November, 1993

5 Dorsch, H., Esch, H.J. and Hensler, P.: 30 Jahre Porsche 911 - Serienmotor (30 years of the Porsche 911 series engine), MTZ 54 (1993), October, 1993

6 Altmann, O., Jeschonnek, P. and Korinth J.: The development of plastic components for engine compartment. Paper given at the Annual Conference on Plastics in Automotive Engineering, Mannheim 1991, VDI Verlag ISBN 3-18-401144-5, pp. 65/105

7 Renaud MC.: (DuPont De Nemours) FE-Studie Ventildeckel, (FE study on valve covers), December, 1992

8 Haldenwanger, H.-G., Mineif, P., Arnegger, K. and Schuler, S.:Kunststoff-Motorbauteile in Ausschmelzkerntechnik am Beispiel eines Saugrohres, (Plastic engine components using the example of an intake pipe), ATZ 89, March, 1987

9 Altmann, O. and Parr, O.: Schmelzkerntechnik im Motorenbau, (Core-melt-out process in engine design), Kunststoffe 81, March, 1991

10 Haldenwanger, H.-G., Leitner, P. and Mineif, P.: Plastic - a material for hybrid construction in modern engine design. Paper given at the Annual Conference on Plastics in Automotive Engineering, Mannheim 1993, VDI Verlag ISBN 3-18-234170-7 pp. 17/104

11 Kohler, H.: Industrial-scale vehicle recycling trial by the German automobil industry. Paper given at the Annual Conference on Plastics in Automotive Engineering, Mannheim 1992, VDI Verlag ISBN 3-18-234165-0 pp. 139/152

12 Schäper, S.: Status report: Polymer recycling with in the German automobil industry′s national trials on scrar vehicle recycling. Paper given at the Annual Conference on Plastics in Automotive Engineering, Mannheim 1993, VDI Verlag ISBN 3-18-234170-7, pp. 227/258

13 Mast, P. and Horsch, G.: Demontage und Verwertung von Kunststoffbauteilen aus Automobilen, (Removal, dismantling and recycling of plastic components from motor-vehicles), FAT-Schriftenreihe 100, April, 1993

Chassis components comprising milled fibre reinforced plastics using the example of the lever, control arm and pedals

Lizhou Yin, Hans-Günther Haldenwanger and Herwig Reim

Development strategy

The technical and scientific development of materials for automotive engineering inevitably results, against the background of increasing component multi-functionality, in hybrid designs which exhibit high levels of component and function integration. For the motor-vehicle development engineer, this demand necessitates the synthesis of the technically possible with the economically feasible. The consistent realization of these product-influencing factors results, in the case of the systematic use of fibre-reinforced plastics, in a functional and rationally priced lightweight system with the overall potential for component weight-savings, and thus greater fuel economy. Despite the numerous evaluation criteria for the development of components and assemblies, heavily loaded members of the suspension, such as the levers, control arm and pedals, etc., can also be the subject of weight-reducing activities. For our purposes, the aluminium components may be taken as the basic cost and weight reference.

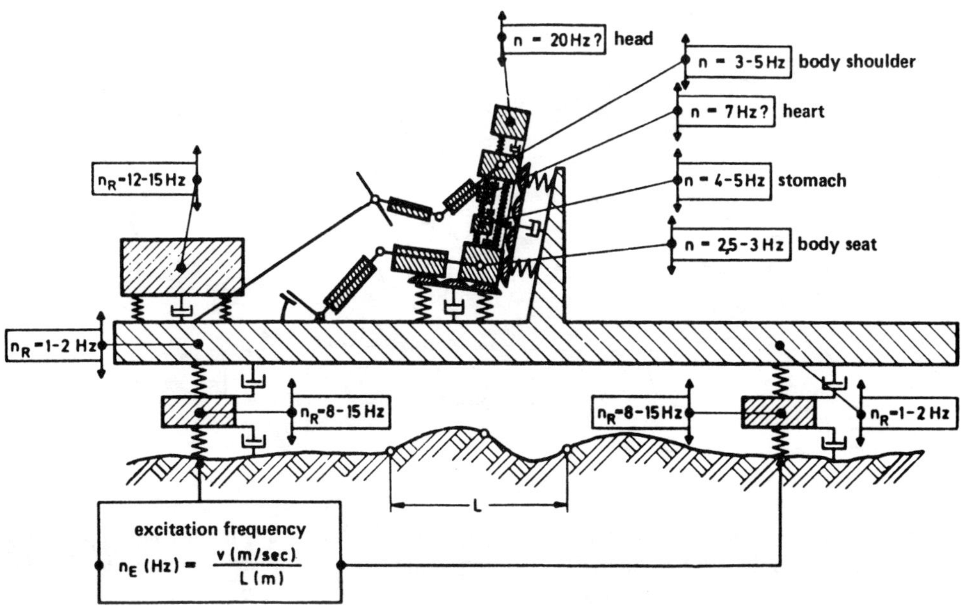

Figure 1 Principle diagram of the equivalent vibration model for the human - seat -vehicle - road system [1].

Requirement profile concepts for pedals and levers

In principle, the well-being of humans in motor-vehicles depends, in addition to psychological factors, i.e., the biological equilibrium of the organic functions in the human body, principally on physical parameters determined essentially by the sitting position and seated posture in the man-seat-vehicle-road system, see Figure 1. The anthropometric design of the seat, the seat position/seat attitude as a function of the positions of the steering wheel and pedals and the support of the relevant parts of the body, the vibrations which are transmitted, and also temperature and humidity at the points of contact with the body (all these are physically quantifiable variables) stimulate in the human organism reactions, the totality of which constitute the well-being of the person involved. These psychological processes in a person are reflected, for instance, in cardiac and muscular activity and, after a prolonged exposure period, become apparent in increased reaction times for the intake and processing of information [1]. It is therefore vital to define the targets for the spatial dimensions of a passenger cell with a high level of user utility at the very earliest phase of vehicle concept definition.

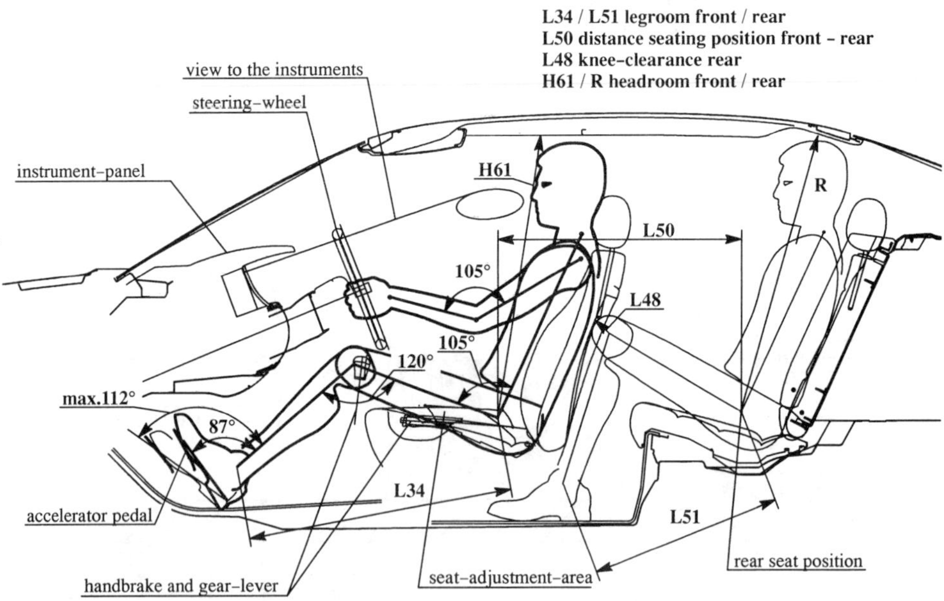

Figure 2 Side view of driver's seat.

The body measurements of potential subsequent vehicle users, and human anthropometric data[1] can be used, in combination with three-dimensional limb-software dummies, to plan the

[1] VDI-Richtlinie 2780 and 2782

seating, and the driver's position as a place of work, with freely selectable dimensions, Figure 2.

Knowledge of body measurements, such as overall dimensions, joint-to-joint measurements and joint angles, together with the relevant mean figures and deviations, can thus be used to state and define the necessary dimensions, incorporating the legal requirements for the driver's place of work for

- steering wheel position
- pedal position (accelerator, brake and clutch pedals)
- seat adjustment range (longitudinal and height adjustment)
- instrument position
- field of vision (forward and lateral)
- rear view, field of mirror vision (internal and external mirrors)
- hand-brake lever position (operating force and angle)
- gear-lever position in all gears

both for the average person and for the 5 % woman and 95 % man, see Figures 3 and 4. Experience gained up to now indicates that the angles (see Figure 2) between the lower abdomen and thigh (105°), the thigh and the calf (120° [112°/118°]), the upper arm and forearm (105° [81°/133°]) produce the lowest level of muscular tension and impairment of venous circulation, particularly from the lower extremities, while avoiding unnecessary pressure (constriction) of internal organs (the stomach) (figures in square brackets are for the

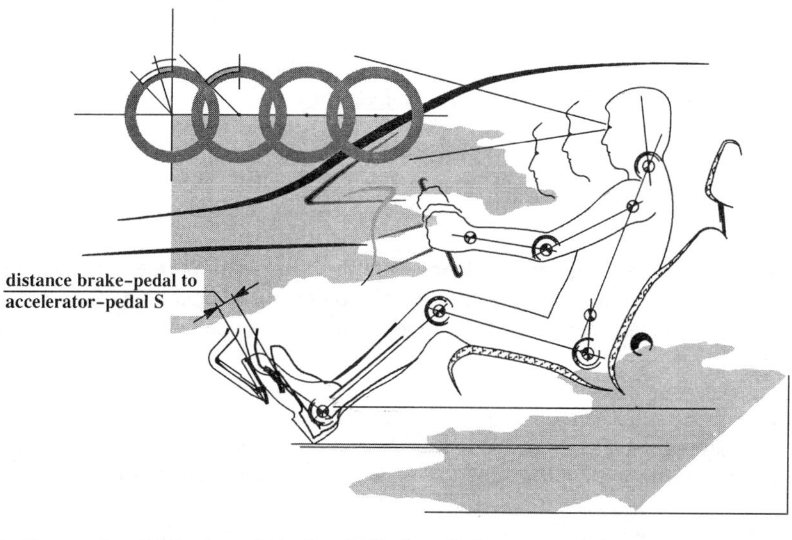

Figure 3 Ergonomics of driver's position (longitudinal section).

5 % woman and 95 % man). This posture is also advantageous when actuation of the steering wheel and the pedals is examined. Despite the fact that the driver's hands and feet demand specific positions on the hand-operated controls and pedals, and that the eyes must be located within a certain zone to permit unobstructed vision in all directions (Eye ellipse, Figure 2) irrespective of body length (position of eyes differs in the longitudinal direction by approx. 144 mm), the potentials for adjustment of the

- seat
- steering wheel (angle and height adjustment)
- steering column (longitudinal and height adjustment)

or even of the pedals (height adjustment) remain a compromise between the breadth of variations of body measurements and their combinations [1,2], see Figure 5.

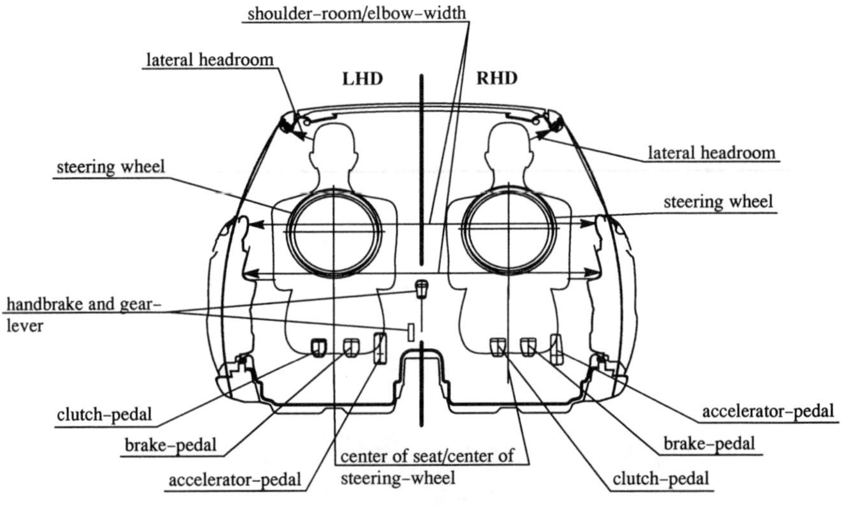

Figure 4 Ergonomics of driver's position (transverse section).

In assessing the effects of a given oscillation, and therefore load, on the multi-mass system "human" (Figure 1), attention must also be directed to the ease of operation of the pedals, and in particular that of the accelerator pedal, as one of the driver's control elements [3], in terms, for instance, of

- vibration exposure (accelerator-pedal tickle) (Figure 6)
- pedal plate arrangement, i.e., overslung or underslung accelerator pedal (Figure 7)
- pedal actuation force as a function of pedal travel (degressive/progressive actuation characteristics),
- height difference between the accelerator and brake pedal plate (Figure 3),
- pedal plate spacing between the accelerator, brake and clutch pedals (Figure 5).

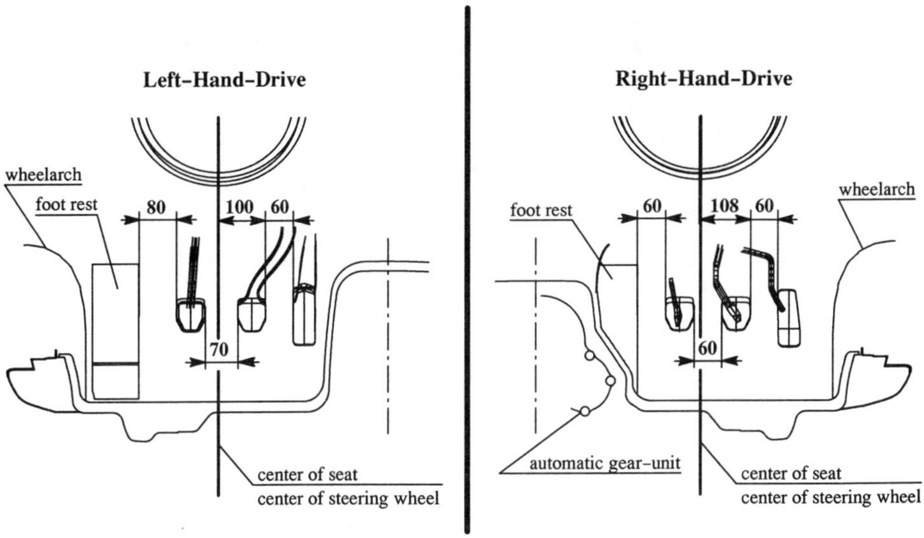

Figure 5 Comparison of pedal arrangement, left-hand drive/right-hand drive.

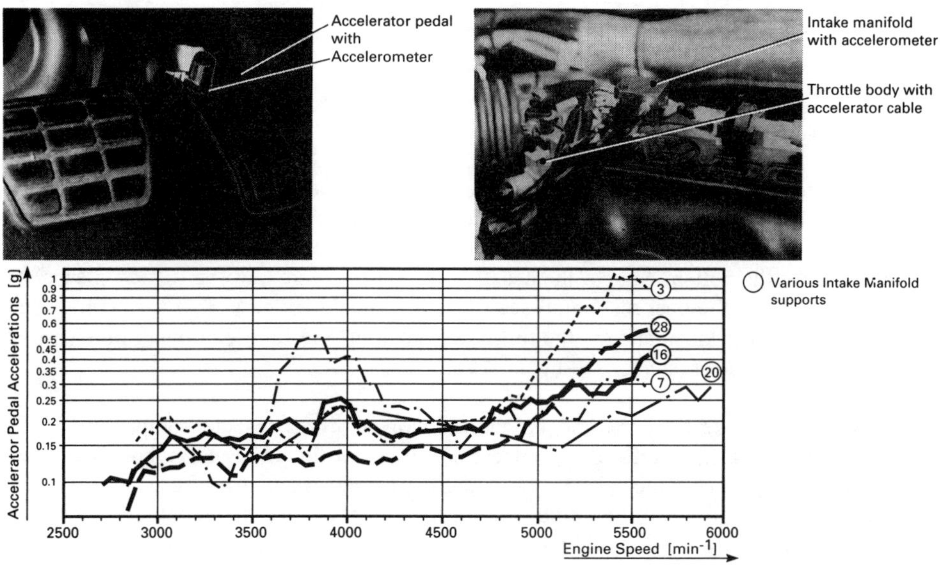

Figure 6 Accelerations in the throttle body/accelerator pedal [3].

For reasons of improved differentiation between the pedals and in terms of the typical modes of pedal use, an overslung arrangement for the brake and clutch pedals is a useful solution, see Figure 8. Ideally, pedal movement should be linear along the path of natural flexing of the leg. For the accelerator pedal, on the other hand, the choice of system between an overslung and an underslung pedal must be examined as a compromise in foot rotary movement around the ankle swivel point with a fulcrum at the heel of the shoe (heel rest point); see Figure 7. Since it is difficult to make allowance in design for the many different sizes of foot, a self-adjusting pivoting pedal plate, see Figure 9, which in a plastics model can be adjusted by means of a beaming hinge, appears recommendable in the case of overslung gas pedals, see Figure 10.

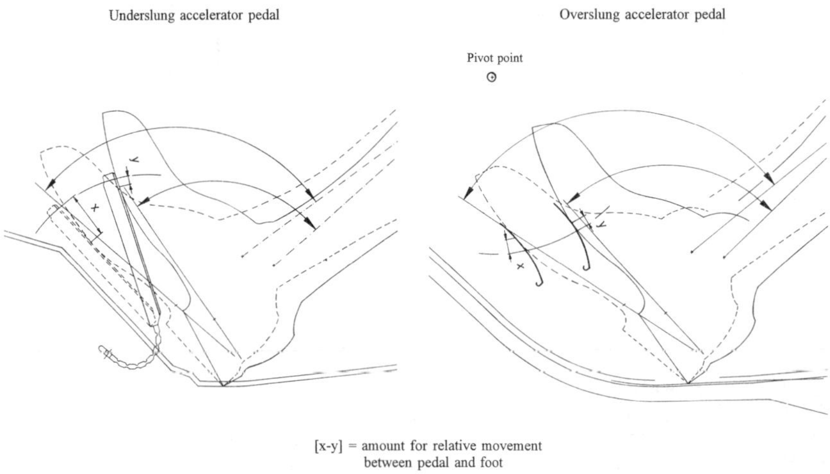

Figure 7 Accelerator pedal variants and their basic kinematics.

Particular attention should be devoted to the accelerator pedal plate's rest position, i.e., although a low accelerator pedal plate position is advantageous for small persons with small feet, the pedal plate must not be set too low for taller drivers with bigger feet (danger of catching against the bottom edge of the pedal!). In addition, there must be sufficient space to permit relocation of the heel rest point, see Figures 7 and 11, in order that it can "track" as a result of movement kinematics in the full power position and for kick-down in the case of automatic transmission systems (see Figure 11).

Examination of the ergonomics and functions involved illustrates the complexity of the requirement profile and, therefore, the concept potentials for the pedals as a whole. An initial approximation, however, indicates good opportunities for the use of cost-effective short-fibre-reinforced thermoplastics and the resultant multi-functional integration potentials which arise from the use of the injection moulding manufacturing process in the case of the accelerator and clutch pedals.

The pedals

Since, as already mentioned, the wide range of requirements made demands a high level of multi-functionality from the pedal mounting assembly, including the steering column and pedal mountings, Figure 12, selection of materials and processes for the individual components should be effected on the basis of the integration potentials for the individual functions.

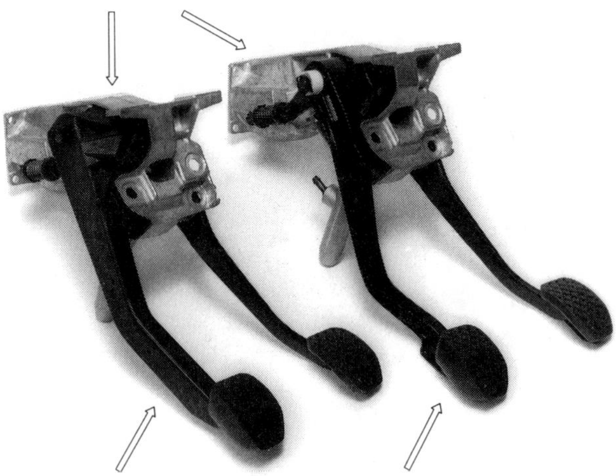

Figure 8 Pedal mounting with overslung pedals.

Accelerator pedal

The example of the accelerator pedal (see Figure 9) may be used to illustrate the multi-functionality via the development path from an accelerator pedal with a fixed pedal plate to an accelerator pedal with a pivoting pedal plate. In an overslung accelerator pedal with a pivoting pedal plate, the translational movement between the pedal plate and the foot is less than with a fixed pedal plate, making it possible to improve ease of operation (see Figures 7 and 11). Vis-à-vis the rigid steel accelerator pedal (Audi 80 B2), with a weight of 400 g (Figure 9), the use of a short-glass-fibre-reinforced plastics accelerator pedal with a pivoting pedal plate (Audi 80 B3, Figure 9) made it possible to achieve weight savings of 43 %. The pedal body in the plastics pedal consists of PA 6.6 GF 35, while the pedal plate with a beaming hinge (Figure 10) is made from polypropylene. The pedal plate and body are joined using an ultrasonic riveting technique.

The weight savings of 172 g also includes the polyacetal bearing support, which as a result of its integrated rotary action plug and snap-connection system for connection to both the pedal body and the pedal mounting, permits significantly easier and quicker installation, see Figure 13.

A further measure of integration to achieve increased function variability can be found in the metal plate, which is located underneath the pedal plate as a balancing weight (see Figure 13). The balancing weight can thus be adjusted systematically and with differing masses to accord with the drive-related oscillation ratios between the throttle and the pedal (Figure 6). Although this physically necessary provision - systematic positioning of additional masses at the most effective point - themselves signify extra weight in the system, this disadvantage can be countered by the function improvement which is achieved, since accelerator pedal tickle is prevented.

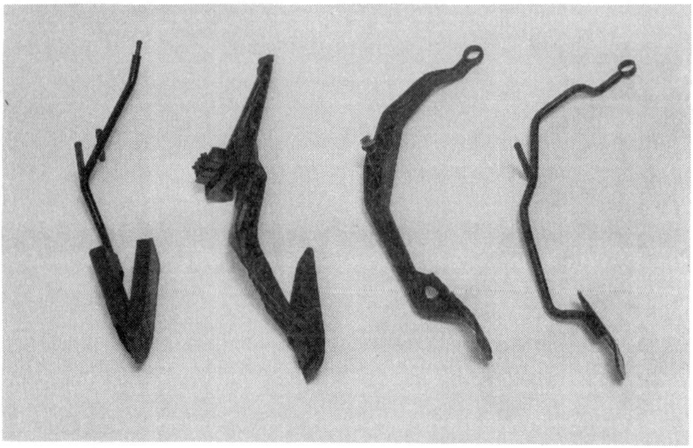

plastics/metal hybrid Pivoting pedal plate 352 g	PA 6.6 GF 35 Pivoting pedal plate 228 g bearing support integrated (40 g)	PA 6.6 GF 35 Fixed pedal plate 186 g	metal Fixed pedal plate 400 g

Figure 9 Accelerator pedal variants (fixed - self-adjusting).

The stiffness and strength requirements for the pedals are subdivided into static strength and stiffness, and strength in service, taking account of the resistance to fluids, temperature, relative air humidity, etc., of thermoplastic materials.

Clutch pedal

The following requirements are made on a plastics clutch pedal with a hydraulic force transmission system:

- Static strength and stiffness in direction of movement

Test force	$F_K = 1000$ N (function force)
Deformation at	$F_K \leq 17$ mm in "dry state"
Deformation at	$F_K \leq 25$ mm in "conditioned state"

- Static strength and stiffness in transverse (to vehicle) direction

Test force	$F_Y = 200$ N
Deformation at	$F_Y \leq 20$ mm in "dry state"
Deformation at	$F_Y \leq 28$ mm in "conditioned state"

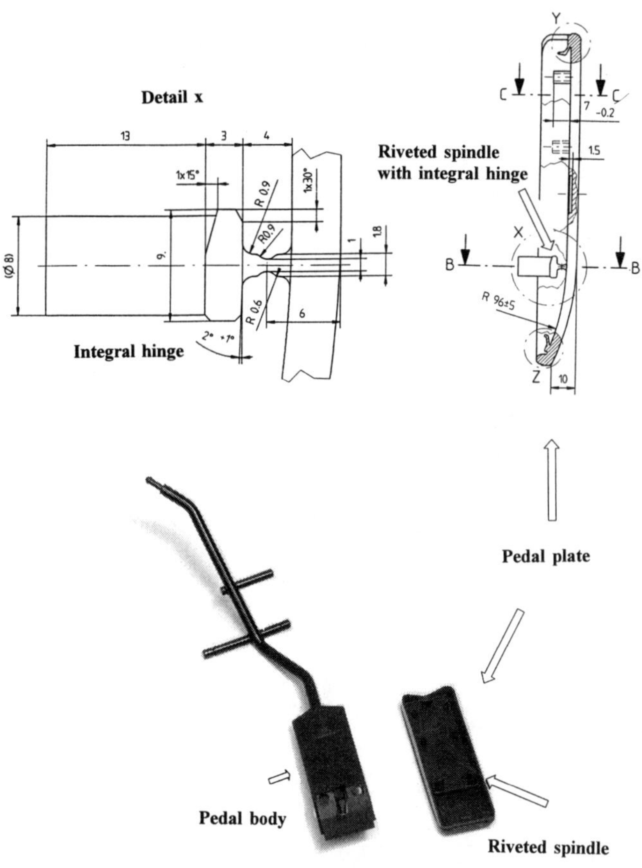

Figure 10 Pivoting pedal with beaming hinge.

It should be noted that the period of exposure to load must be smaller than or equal to 5 seconds for both directions of force action and that the measured data must be determined within 5 seconds.

The term "dry state" signifies prior storage at +80 °C for a period of 72 hours, while "conditioned state" signifies prior storage at +55 °C and not less than 95 % relative air humidity for a period of 48 hours.

- Strength in service

Actuation force	F_K from 0 to 350 N
Frequency of actuation	≤ 3 Hz
Mean cycles	$\geq 200\,000$
Standard deviation	$s_{log} \leq 0.25$

These tests of dynamic strength were performed for plastics component developments at temperatures of - 40 °C, room temperature, + 60 °C and + 80 °C, in each case with prior exposure to fluctuating temperatures and climatic conditions for 96 hours[2].

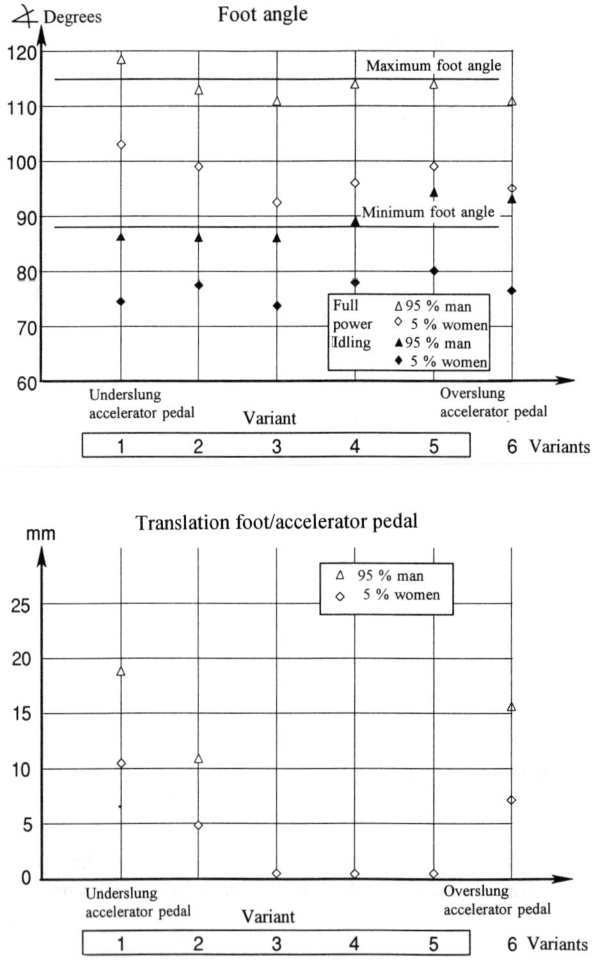

Figure 11 Foot angles and translation between foot and pedal.

The test apparatus, Figure 14, shows that actuation force F_K applies both bending and torsional load to the clutch pedal. The section of torsional and bending-load exposed levers optimized for stiffness and weight results in a closed hollow section and will need to be designed more rectangular or more circular or elliptical, depending on whether bending or torsion loads dominate [4], see Figure 15.

[2] Test procedure PV 1200 (Audi)

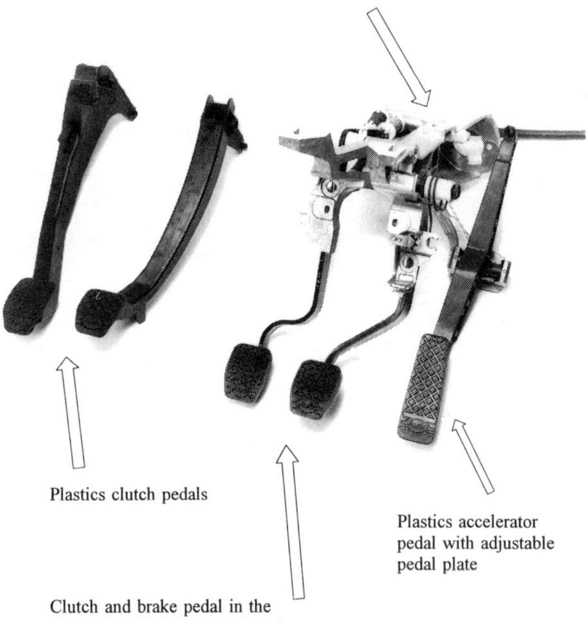

Figure 12 Audi S 2 pedals; overslung type

An easy-to-use Finite Element Program[5] is applied to perform preliminary dimensioning for bending load on the basis of stiffness, in the form of a maximum acceptable deflection at corresponding clutch actuation loads, following definition of the hollow-box section which is weight-optimum under the given circumstances. If the available space is utilized, the moment of plane area, and therefore flexural strength, are influenced by the wall thickness of the hollow section.

The following applies:

$$\textit{Flexural strength} \quad E \cdot I = \frac{M_b \cdot e}{\epsilon_{per}} \tag{1}$$

The second variable which is definitive for flexural strength is the modulus of elasticity of the pedal material. Variation of modulus of elasticity and moment of plane area can be used, making allowance for maximum permissible strains, to calculate the pedal deformation distances under loads in accordance with the force diagram, (see Figure 16). The result of this

[5]FINAL, from BAYER AG

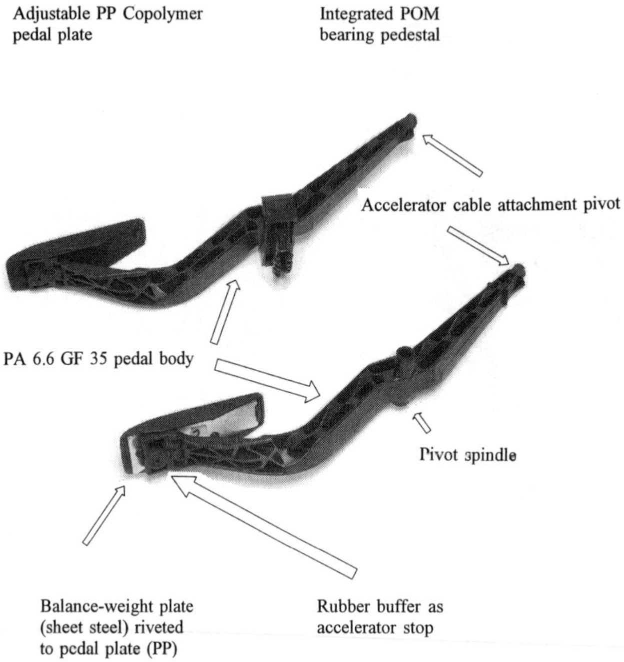

Figure 13 Glass-fibre-reinforced polyamide accelerator pedals.

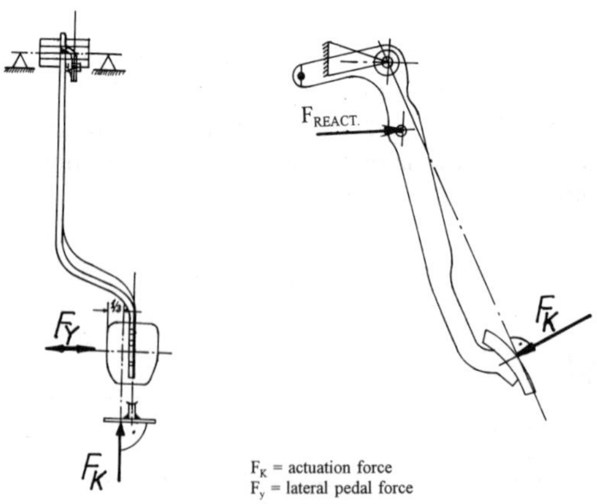

Figure 14 Test apparatus for clutch pedals[3].

[3] TL-VW 82183 (Audi)

Section No.	Section geometry		Assessment factors						Quality index
			Stiffness			Load capacity			
			Tz	By	Bx	Tz	By	Bx	
1	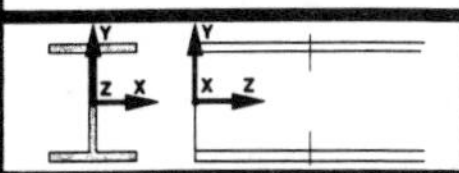		1	1,0	1,0	1,0	1,0	1,0	XII.
2	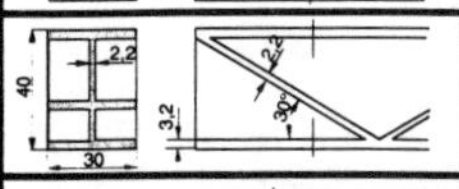	A	8	1,1	1,1	1,6	1,2	1,1	XI.
		B	7	0,9	0,9	1,3	1,0	0,9	
3	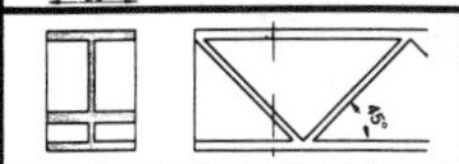	A	13	1,0	1,1	2,1	1,1	1,1	X.
		B	10	0,8	0,9	1,6	0,8	0,9	
4	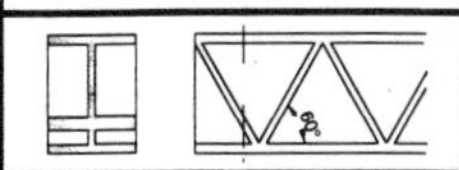	A	15	1,0	1,1	2,5	1,0	1,0	IX.
		B	11	0,7	0,8	1,8	0,7	0,7	
5	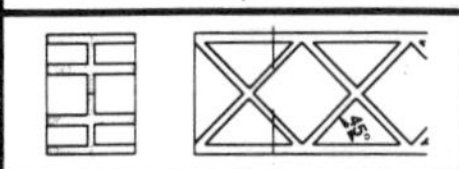	A	28	1,1	1,1	4,7	1,1	1,1	V.
		B	18	0,7	0,7	3,0	0,7	0,7	
6	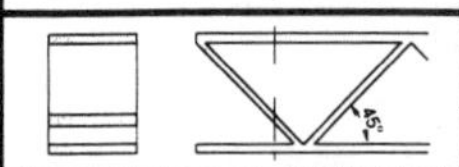	A	13	1,0	0,8	2,1	1,1	0,8	VI.
		B	14	1,1	0,9	2,3	1,2	0,9	
7	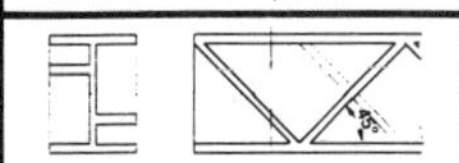	A	14	1,0	1,0	2,1	0,9	1,0	VIII.
		B	12	0,9	0,9	1,8	0,8	0,9	
8	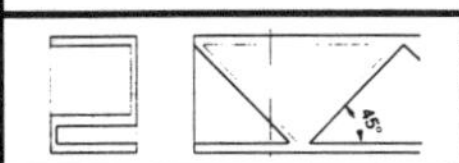	A	35	1,7	1,0	6,6	1,4	1,0	III.
		B	27	1,3	0,8	5,1	1,1	0,5	
9	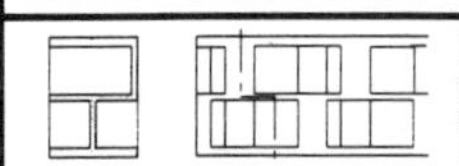	A	16	2,6	1,1	4,1	1,7	0,6	VII.
		B	12	2,0	0,9	3,2	1,3	0,5	
10	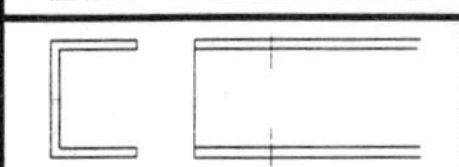	A	1	1,9	1,0	1,0	1,4	1,0	XIII.
		B	1	1,9	1,0	1,0	1,4	1,0	
11	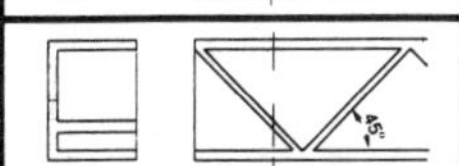	A	32	2,0	1,0	3,3	1,5	1,0	IV.
		B	25	1,6	0,8	2,6	1,2	0,8	
12	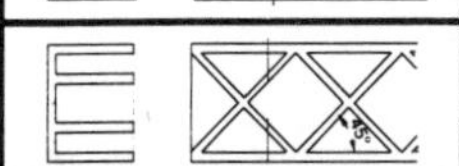	A	54	2,0	1,0	6,6	1,6	1,1	II.
		B	35	1,3	0,7	4,2	1,0	0,7	
13	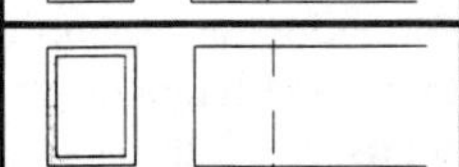	A	73	3,7	1,1	13,5	3,8	1,1	I.
		B	54	2,7	0,8	10,0	2,8	0,8	

Stiffness: Resistance to deformation
Load capacity: Load required to cause fracture
Factors "A": (on dark background) referred to Section No. 1
Factors "B": as for "A", but referred to volume

All figures relate solely to the geometrical ratios
Quality index: Order of specific (volume-based) torsional stiffness (factors B of Tz)

Figure 15 Comparison of stiffness in sections[4]

[4] Source: BAYER, ATI 786

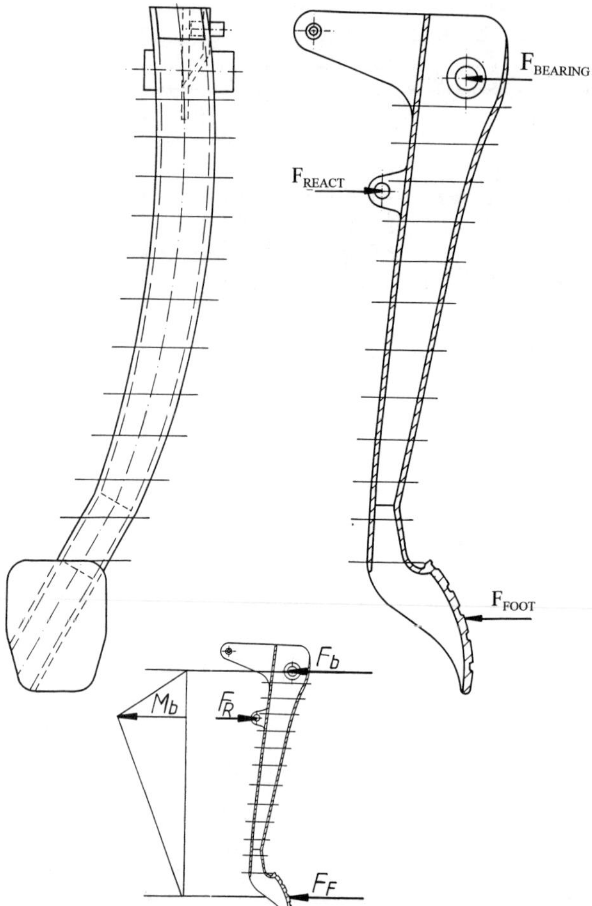

Figure 16 Force diagram for clutch pedal for FEM calculation.

preliminary dimensioning operation provides on the one hand information on feasibility and, on the other hand, indicates the geometrical relationships and narrows the range of materials which can be selected. After careful consideration and comparison of the mechanical properties of short-fibre-reinforced engineering thermoplastics, such as strength, modulus of elasticity, impact strength and reactions to temperature, time and fluids, Audi initially decided in favour of the glass-fibre-reinforced polyamides, and ultimately concentrated on PA 6.6 GF 35 and PA 6/6T GF, see Figure 17 and Table 1. For the purpose of preliminary dimensioning, the materials behaviour of these two polyamide materials can be assumed to be linear isotropic for short-term exposure to loads and linked to a secant modulus of 6500 N/mm^2. After FEM dimensioning and selection of material, the construction of a so-

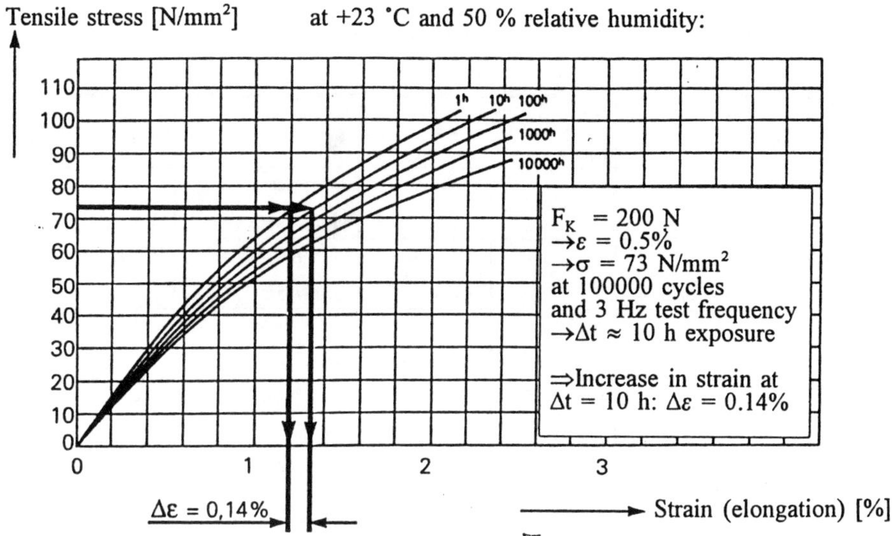

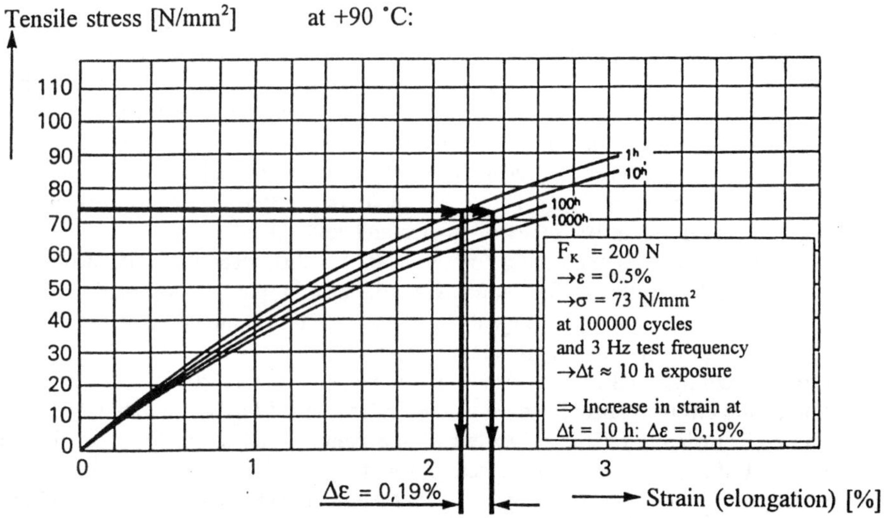

Short-term strength at +23 °C and 50 % relative humidity, test speed = 2mm/min:

Tensile stress σ_s = 147 N/mm²

Strain (elongation) at tensile stress ε_s = 3 %

Figure 17 Isochronous stress-strain diagrams for PA 6.6 GF 35 at room temperature and 50% relative humidity and at +90 °C[6] .

[6]BASF data sheet, DIN 53444

Table 1 Materials appraisal for clutch pedal C4 (by Riesselmann & Sohn)

Material	Test condition in accordance with TL 82 183	Deflection at 500 N [mm]	Deflection at 1000 N [mm]	Fracture load [N]	Moisture absorption [%]	Weight, calculated [gr.]
Polyamid 6.6 GF35 Ultramid A3WG7 **)	dry	10,9 +/- 0,6	20,9 +/- 0,8	1446 +/- 261	1,62	360
	conditioned	13,2 +/- 0,2	27,4 +/- 0,5	1459 +/- 143		
Polyamid 6.6 GF60 Zytel EFE 7220 *)	dry	7,5 +/- 1,3	14,8 +/- 0	1309 +/- 0	1,76	445
	conditioned	8,3 +/- 1,9	16,1 +/- 1,27	1380 +/- 252		
Polyamid 6.6 GF50 Ultramid A3WG10 **)	dry	8,3 +/- 0,2	15,9 +/- 0,3	1333 +/- 160	1,43	400
	conditioned	9,7 +/- 0,7	19,9 +/- 0,8	1594 +/- 88		
Polyamid 6/6T GF35 Ultramid T KR 4355 G7 **)	dry	10,7 +/- 0,4	20,0 +/- 0,6	1699 +/- 200	0,90	366
	conditioned	10,3 +/- 0,2	20,1 +/- 0,2	1989 +/- 32		
Polyamid 6/6T GF50 Ultramid T KR 4355 G10 **)	dry	7,4 +/- 0,8	13,8 +/- 0,8	1167 +/- 341	0,69	400
	conditioned	6,9 +/- 0,1	13,3 +/- 0,2	1774 +/- 153		

*) Number of test objects = 2 each
**) Number of test objects = 5 each

called "resin model" which accords with the latest design status is recommendable for the purpose of identification of critical areas on the basis of model laws (similarity observation) [4]. This first clutch pedal prototype is produced from a resin (e.g. epoxy resin), using which the stiffness of PA 6.6 GF 35 can be derived. This system cannot be used for assessment of strength. The clutch pedal prototype for evaluation of stiffness is subsequently used to illustrate the stress distribution in the resin clutch pedal when exposed to load by means of holographic interferometry, see Figure 18 [4]. The holographic interferograms generated for the pedal provide information on zones of higher and lower loading, see Figure 19. The macroscopic interference lines mark the deformation state and connect all points of the same deformation amplitude on the surface. Sudden variations in the plot or density of the interference lines indicate an abrupt change in deformation and thus increased material effort. After completion of these measurements, strain gauges can be applied to the areas of interest, i.e., those of elevated material effort, of the resin model, see Figure 20, in order to permit identification of the effort as specific strain values under defined loads, see Table 2. All this information and knowledge on the behaviour of the clutch pedal under load can then be followed by an FEM analysis extremely effective from the inception, if this is still considered necessary. Modern FE programs make it possible, inter alia, to take account even of nonlinear materials behaviour (Figure 17), and thus to calculate deformation behaviour into ranges very close to total failure.

Figure 21 shows the component result for the development cycle described. Such systematic preliminary development provides proof that complex-geometry hollow bodies can be not only dimensioned and calculated but also produced using thermoplastic injection moulding methods to achieve weight advantages. The producer must possess a significant know-how potential in the design and support of long slim rotatable cores, however, if internal pressure injection moulding technology is rejected for weight reasons.

Testing of the plastics clutch pedal for the Audi 100 for strength, stiffness and pedal bearing design was accomplished without difficulty. The bearing design was tested in advance on the basis of preliminary tests for suitable material combinations for the metal bearing spindle and the integrated plastics bearing shell (glass-fibre-reinforced polyamide) for bearing wear and bearing noise. The appropriate material combinations for the glass-fibre-reinforced polyamide bearing shell is a bearing spindle consisting of nitride-hardened steel with a surface hardness of 800 HV 10 and a peak-to-valley roughness of $R_z = 2.5$ μm.

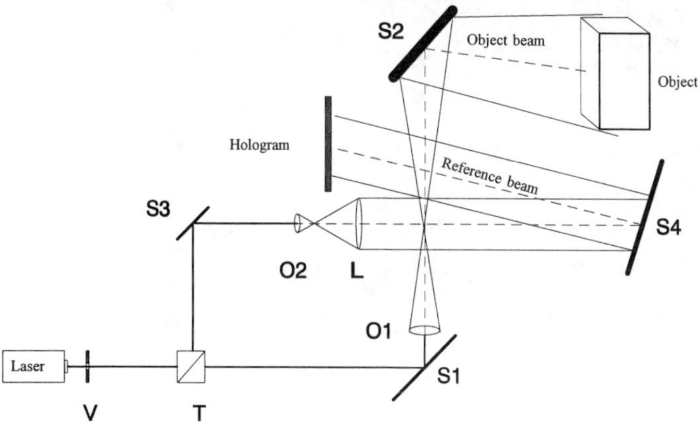

Figure 18 Holographic interferometry; path of the rays[7].

At Audi, series-production adoption of the lightweight plastics clutch pedal was ultimately rejected as a result of the vibration behaviour of one unit variant, i.e., when the pedal is pressed at an engine speed of approx. 4000 revolutions per minute, vibrations are transmitted via the hydraulic system into the pedal body, and result in tangible pedal vibration. The damping provisions necessary in accordance with:

$$f_e = \frac{1}{2\pi} \cdot \sqrt{\frac{c}{m}} \tag{2}$$

generally necessitate additional weights, however, which then negate the weight advantage of 21 % otherwise obtained over the aluminium pedal.

The same applies to the potential cost advantage initially achieved over the aluminium pedal, i.e., this advantage, too, may be completely negated or even reversed, depending on the scope of the vibration damping provisions necessary.

Under the circumstances discussed above, a plastics clutch pedal at present produces neither a weight nor a cost advantage over a die-cast aluminium type in the Audi range, with the result that the positive preliminary development phase is currently not capable of justifying series production. One must, however, assume in principle that a glass-fibre-reinforced

[7]Source: BASF, Ludwigshafen

polyamide clutch pedal could achieve savings of up to 25 % in weight and approx. 10 % in costs vis-à-vis an aluminium clutch pedal under other acoustic boundary conditions.

Hollow-box section:

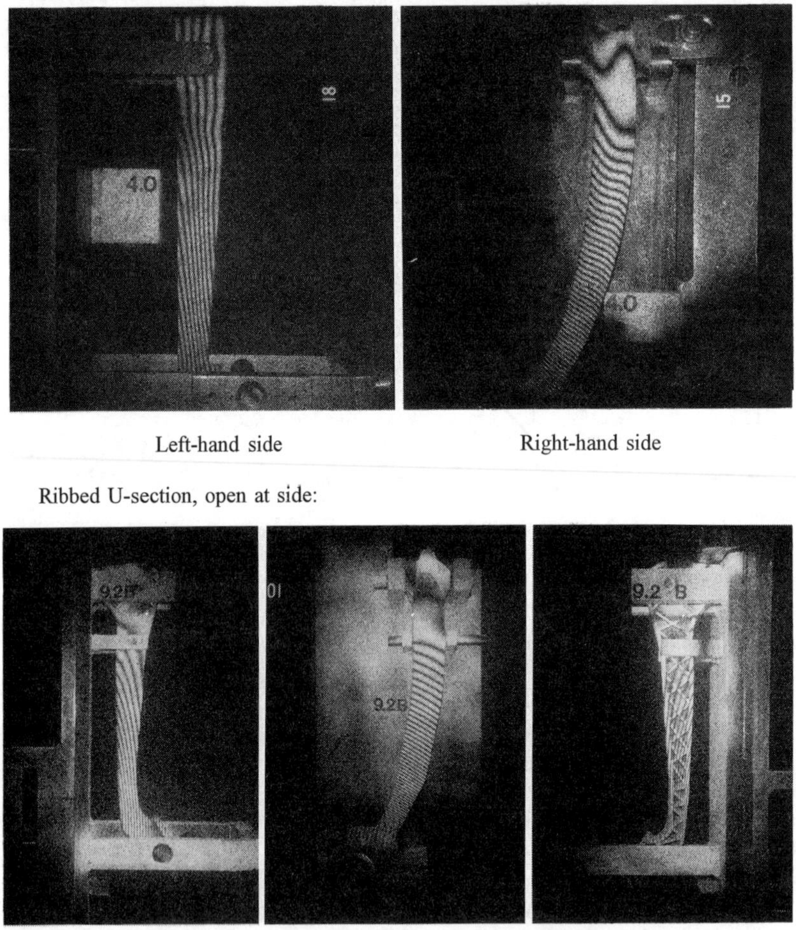

Left-hand side | Right-hand side

Ribbed U-section, open at side:

Left-hand side | Front view | Right-hand side

Figure 19 Holographic interferograms; loads on resin clutch pedal models in the direction of actuation[7].

The functional disadvantages specific to plastics, such as the fact that "bending is indicated in the form of cracking, and thus loss of stiffness, under exposure to excessive load" - will probably preclude the use of short-glass-fibre-reinforced plastic materials for brake pedals. A similar situation also pertains in the case of the multi-functional pedal mounting (i.e., crash-compatible steering-column fixings and Audi Procon-Ten safety steering system [4]).

Clutch pedal, left-hand side

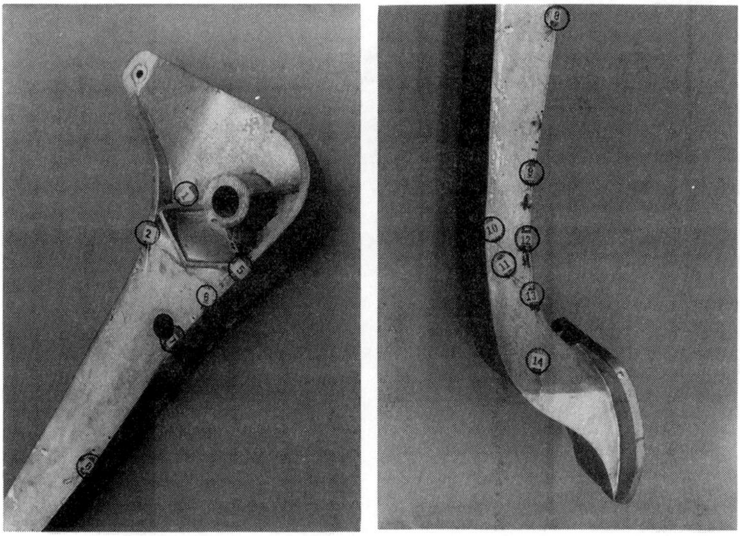

Upper section Lower section

Figure 20 Resin model showing strain gauges[7]

Table 2 Strain measurements in a resin model[7]

Direction of load	from front in direction of actuation	from left
Pedal force	200 N at the left-hand third point of the pedal plate at angle of 5°	100 N
Strain gauge No.	Strain in [‰]	
1	0,05	1,97
2	- 0,31	0,84
3	0,70	1,52
4	- 0,46	0,86
5	0,78	2,90
6	0,28	1,30
7	5,00	2,05
8	2,50	0,68
9	2,30	0,58
10	- 1,00	1,10
11	- 0,18	1,27
12	2,45	0,66
13	2,50	0,44
14	0,30	0,22

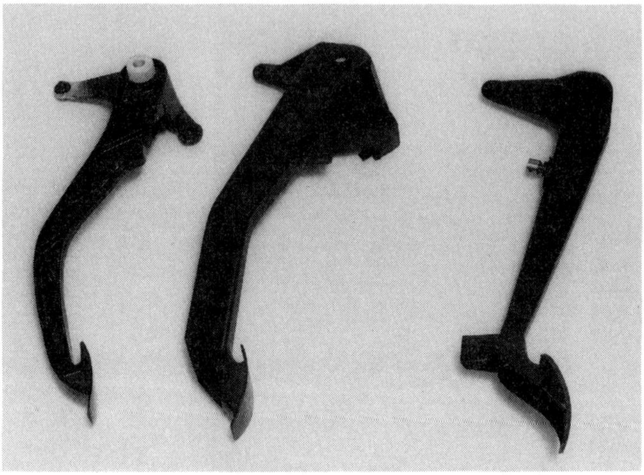

Aluminium	PA 6/6T GF 35	PA 6.6 GF 35
C type	for C type	for B type
448 g	352 g	236 g

Figure 21 Plastics and aluminium clutch pedals.

Levers, arms and connecting rods

Independent suspension systems are now used predominantly for both the front and rear suspension of modern cars, thanks to the advantages they offer in terms of:

- space economy
- lower weight
- optimum kinematics, since no cross-influencing between the wheels and use of the multi-arm levels into individual arms with individual pivot points is possible,

In these multi-arm axle types, such as the

- double-wishbone axle
- wishbone axle
- trailing link (multi-link suspension) axle

aluminium alloys and fibre-composite plastics are also potential problem-solvers for multi-arm levels in individual levers, arms or connecting rods as compared to conventional steel materials [4], see Figure 22.

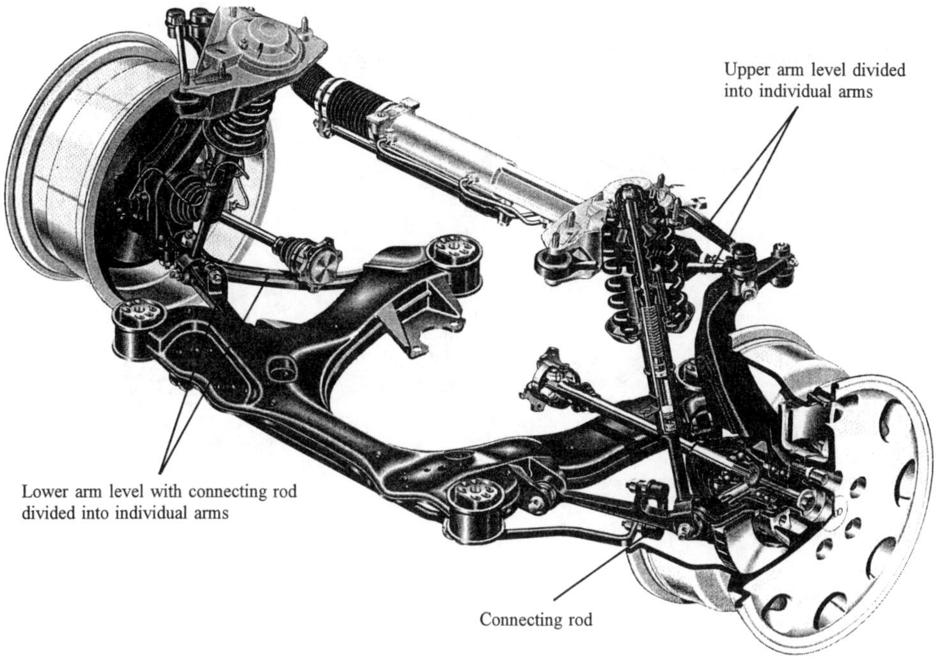

Figure 22 Front axle showing multi-arm levels.

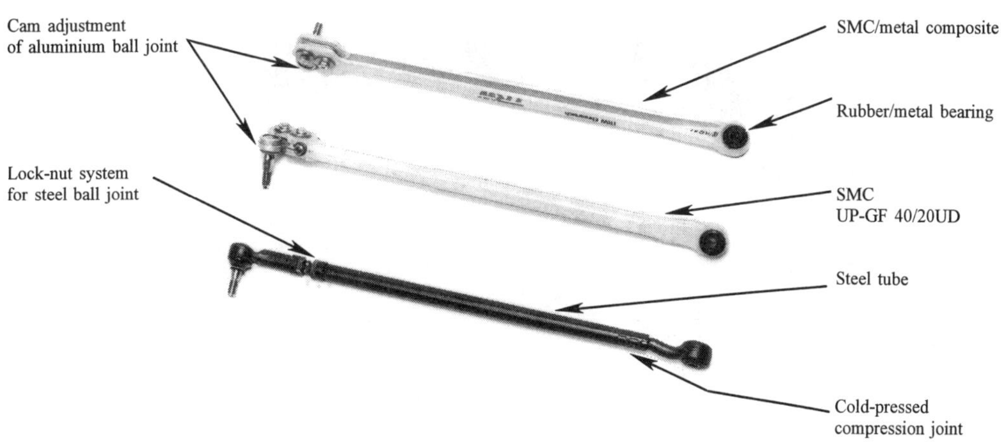

Figure 23 Track rods in various materials.

These heavily loaded arms, equipped with additional bearing points as a result of lateral and longitudinal wheel forces, can essentially be attributed in the points of force transmission to exposure to compression-tensile loads, with the result that even low-price thermoplastic injection moulding materials with short-glass-fibre-reinforcement can be used.

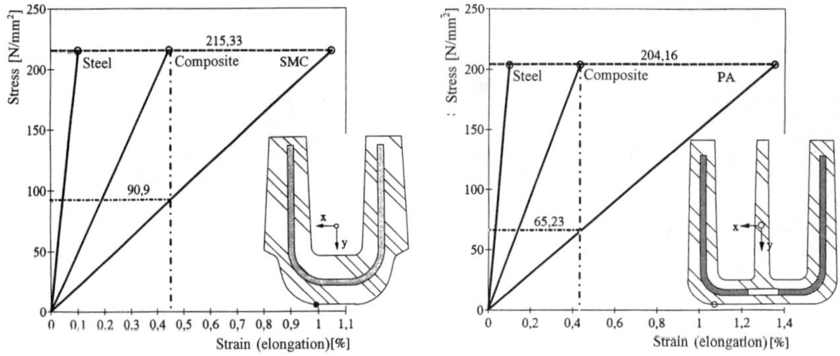

Estimation of stresses on the SMC/metal composite track rod

Figure 24 Steel/plastics composite track rod.

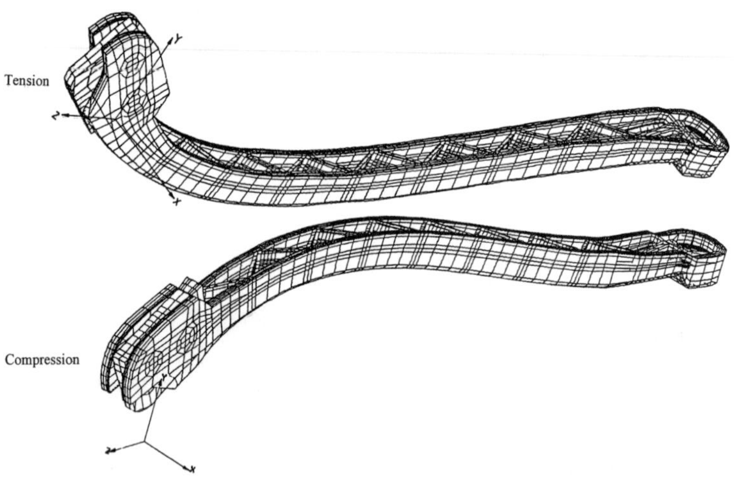

Figure 25 Track rod deformations.

Track rod

If directionally glass-fibre-reinforced SMC (UP-GF 40/20) is used, weight-savings of approx. 43 % can be achieved as compared to the standard steel version (steel tube with bolted or cold-pressed forged steel joints in the track rod of the Audi 100, for instance, see Figure 23. Since this type does not meet the statical misuse loads of tensile force ≥ 28 kN and compressive force ≥ 14 kN with an adequate safety margin, however, a U-section insert of deep-drawn steel sheet can be used as a reinforcement for the SMC, see Figure 24.

Table 3 Comparison of results of FEM calculations for the various track rods

	Load case Tension 28 kN Compression 14kN Oper. load 9,5 kN	Weight [g]	σ_{v1} [N/mm^2]	Safety margin against fracture	σ_{v2} [N/mm^2]	Maximum shift [mm]	Shift in force action [mm]
SMC track rod	Tension	825	220	1	160	9.04	3,0
	Compression		110	1,19	78	4,52	1,5
	Operating load		74		54	3,1	1,02
SMC/metal composite track rod	Tension	1018	120	1,83	99	4,7	1,6
	Compression		58	2,81	47	2,28	0,7
	Operating load		41		33,6	1,6	0,6
Plastics/metal composite track rod with no ribbing	Tension	827	95	2,21	68	4,59	1,52
	Compression		46	2,38	33	2,30	0,775
	Operating load		32		23	1,56	0,6
Plastics/metal composite track rod with a longitudinal rib	Tension	861	95	2,21	68	4,44	1,50
	Compression		46	2,38	33	2,22	0,75
	Operating load		32		23	1,5	0,50
Plastics/metal composite track rod with 45° ribbing	Tension	867	95	2,21	68	3,93	1,5
	Compression		46	2,38	33	1,96	0,75
	Operating load		32		23	1,33	0,5
Plastics/metal composite track rod with ribbing at 30°	Tension	862	95	2,21	68	3,77	1,5
	Compression		46	2,38	33	1,88	0,7
	Operating load		32		23	1,28	0,5

Amortization of extra vehicle price [A] by fuel cost economies

$$A = \frac{k \cdot m \cdot d}{100} \quad [\mathrm{DM}]$$

k = fuel costs [DM/l]
m = consumption decrease achieved [l/100 km]
d = Amortization mileage [km]

Lightweight construction costs divided by consumption cut

$$t = \frac{k \cdot d}{100} \left[\frac{\frac{\mathrm{DM}}{\mathrm{l}}}{100\ \mathrm{km}}\right]$$

k = fuel costs [DM/l]
d = Amortization mileage [km]

Example: $t = \frac{1{,}5 \cdot 45\,000}{100} = 675 \left[\frac{\frac{\mathrm{DM}}{\mathrm{l}}}{100\ \mathrm{km}}\right]$

Marginal lightweight construction costs (LGK) with influence of weight

$$LGK = \frac{c \cdot t}{100 \cdot p} \left[\frac{\mathrm{DM}}{\mathrm{kg}}\right]$$

$$p = \frac{\text{Price}}{\text{Costs}} \approx 1{,}5 \text{ (Variable)}$$

c = Weight influence on consumption $\frac{0{,}6\ \mathrm{l}}{10^4\ \mathrm{km\ kg}}$

t = Lightweight costs divided by cut in consumption $\left[\frac{\frac{\mathrm{DM}}{\mathrm{l}}}{100\ \mathrm{km}}\right]$

Figure 26 Marginal costing for lightweight construction.

Approximate dimensioning indicates that weight-savings of around 28 % as compared to the standard version can be achieved, despite the steel insert. As a further FEM analysis proves, both the stress level and weight can be reduced, by approx. 15 % and approx. 28 %, respectively, see Figure 25, by means of the use of a thermoplastic injection moulding version and U-section ribs, compared to the SMC/metal composite type, signifying that weight savings of 1.2 kg per vehicle would be possible. Although the technical feasibility of the composite fibre/plastics track rod as an alternative to the conventional steel type has been demonstrated at Audi [5] both by means of directionally long-glass-fibre-reinforced SMC press moulding and by means of a short-glass-fibre-reinforced thermoplastic injection moulded model with a metal reinforcement, see Table 3, the economic advantage for the various types of plastics is not convincingly demonstrated (thermoplastic injection moulded part probably "neutral" in terms of component costs, but with higher investment costs than for steel), despite a lightweight bonus [4] - extra costs for cost-effective lightweight construction 3.25 DM/kg of weight saving, see Figures 26 and 27, with the result that series production cannot be recommended. It should be noted in this context that calculation of the equivalent modulus of elasticity cannot be dispensed with in the dimensioning of metal/plastics composites. The following relationship applies universally:

$$E_{mix} = \frac{A_{inlay}}{A_{total}} \cdot E_{inlay} + \frac{A_{plastics}}{A_{total}} \cdot E_{plastics} \tag{3}$$

$$\epsilon_{mix} = \frac{\sigma_z}{E_{mix}} \tag{4}$$

$$\epsilon_{mix} \cdot \frac{215.33}{48546} \cdot 100\ \% = .44\ \%$$

Stabilizing connecting rod

Torsion bar stabilizers are frequently used to modify the roll-steer behaviour of a vehicle. Road behaviour can be adjusted by means of a stronger stabilizer at the front to give more understeer and with a stronger stabilizer at the rear to give more oversteer.

The stabilizers are sometimes connected directly to the wheel carriers or supporting arms (Figure 22), but increasingly often via a connecting rod which is usually equipped with two ball-joints or a rubber bearing combination, see Figure 28. The ball-joints and rubber bearings transmit the forces which are generated upon the occurrence of one-sided jounce and rebound of the wheels into the connecting rod in the form of tensile and compressive forces. Since these rods form part of the unsprung mass of the vehicle, a reduction of their weight is of great significance not only for fuel consumption but also for ride-comfort and strength in service ($F = m \cdot b$).

The design of a stabilizing connecting rod is defined primarily by axle kinematics and the space available. It proved possible to convert the connecting rods for the Audi 80 model to thermoplastic injection moulding reinforced with short-glass-fibre within the scope of a cost and weight reduction program; these types have been in series production since October, 1993, see Figure 29.

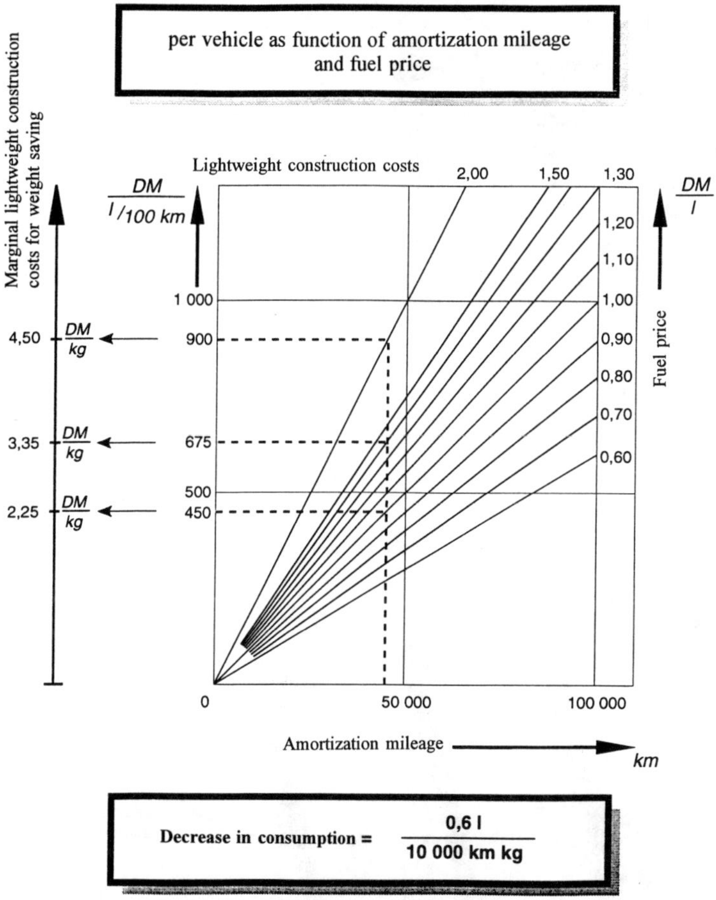

Figure 27 Marginal costs for lightweight construction.

The test program for this type of connecting rods presupposes the following parameters:

Operating temperature	T = 70 °C
Static loads	$F_{tensile}$ = 5 kN, $F_{compressive}$ = 5.5 kN
Dynamic loads	
Single stage test	F = ± 2.2 kN, N = 3 x 10^5 cycles
Operating load test	see collective load group, Figure 30.

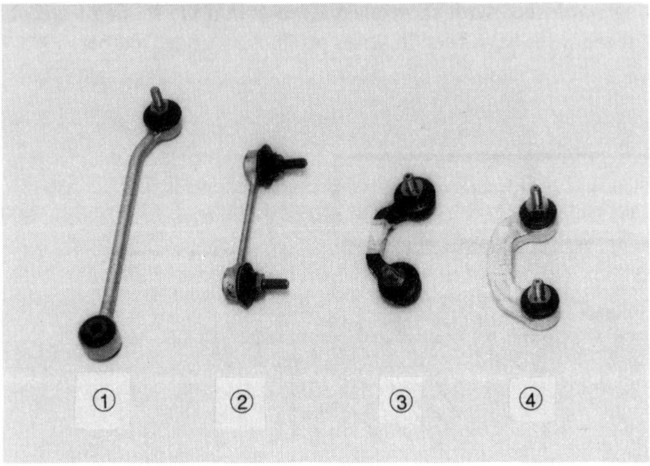

① Audi 80, rear axle
② Audi 100, rear axle
③ Audi A8, front axle (steel type)
④ Audi A8, front axle (aluminium type)

Figure 28 Connecting rods for Audi vehicles.

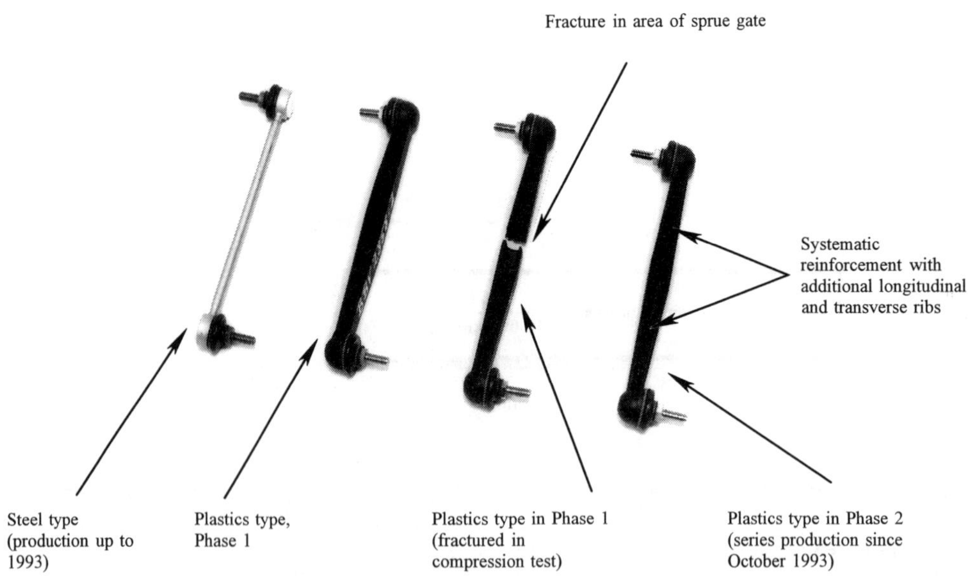

Figure 29 Connecting rod, Audi 80.

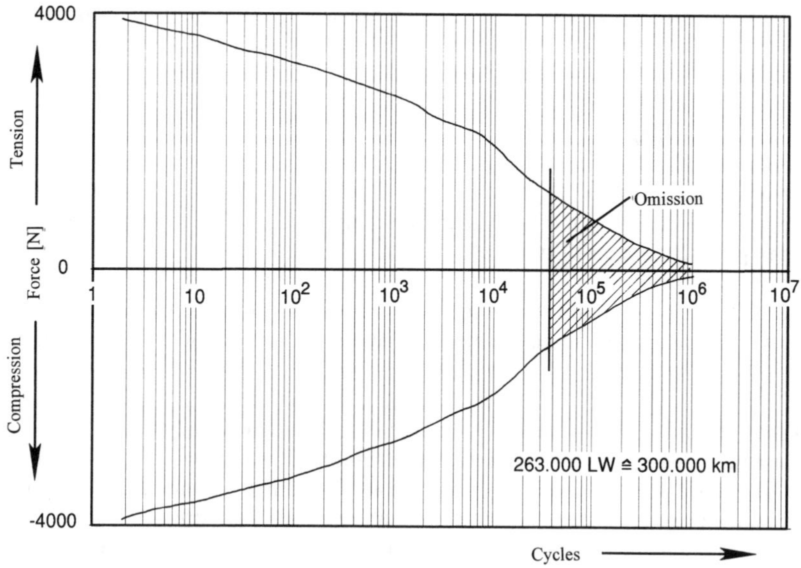

Figure 30 Combined loads for connecting rod, Audi 80.

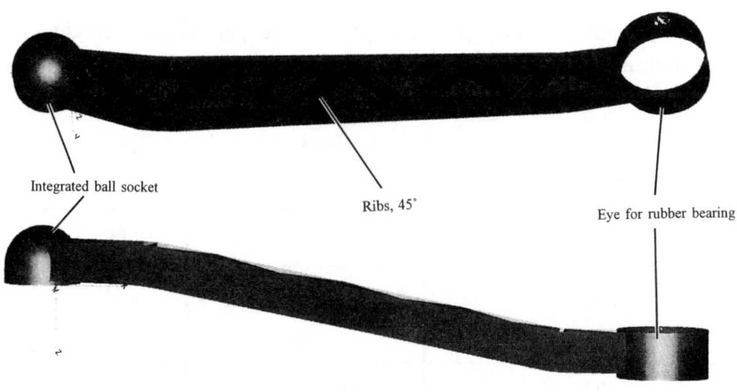

Figure 31 Dropped connecting rod.

Since this is a component which is subject purely to tensile-compressive loads, FEM analysis is not absolutely necessary. Following a strength analysis in conjunction with the Euler's critical buckling load case - both ends flexibly supported - it became apparent that the material needs to have a tensile strength of at least 60 MPa, a fatigue strength of not less than 26 MPa and a modulus of elasticity of not less than 6400 MPa at + 70 °C in condition state in the cross section (Phase 1); see Figure 29. These figures are, therefore, in the range achievable by the majority of glass-fibre-reinforced thermoplastics. The remaining component-specific requirement criteria, such as

- impact toughness at - 40 °C
- sensitivity to stress cracking
- resistance to fluids
- cost-effectiveness
- component quality assurance by means of manufacturing process quality assurance
- large-run series production capability

are adequately fulfilled by polyamide reinforced with short glass fibres (PA66-GF43), with the result that this material appears to be the most suitable for complex and expensive series development.

Despite careful strength and stiffness calculation, including a rheological study using materials characteristics data provided by the raw materials manufacturers, it was apparent in the critical buckling load test (Phase 1, see Figure 29), that the more isotropic component structure exhibits a significant fall in its materials characteristics data and, therefore, a stiffness step (softer) as a result of random fibre orientation in the heavily load-exposed sprue gate area at the mid-point of the connecting rod at + 70 °C and 2.5 % water content, inevitably resulting in failure of the component (Figure 29). Empirical determination of the critical strain in the endangered cross-section (first damage in the σ-ε-diagram) provides information on the need for increasing the moment of plane area by means of additional ribs.

Both the results of this component geometry (Phase 2, see Figure 29) and the weight and component-cost savings permit the deduction that coaxially tensile/compression loaded connecting rods in PA66-GF43 are suitable for large-run series use, see Table 4.

Table 4 Potentials for weight and cost savings.

Connecting rod	Weight [g/rod]	Saving* [g/vehicle]	Cost factor
Steel connecting rod (series production to Oct. 1993	369	0	1
Plastics connecting rod (Phase 1)	253	232	
Plastics connecting rod (Phase 2, series production since Oct. 1993	272	194	0,87

* as compared to steel connecting rod

Dropped connecting rod

If the shear centre of a ribbed U-section can be attributed at relatively low loads on a connecting rod to a coaxial load in a component plane, it may be made elbowed from short-glass-fibre-reinforced thermoplastic injection moulding in the plane offset at 90°, see Figure 31. If, in addition, the ball socket for the ball-jointed spindle bearing is integrated, the weight-savings as compared to steel can be roughly estimated at approx. 48%. The multi-functionality described also makes it possible to predict a component-cost reduction.

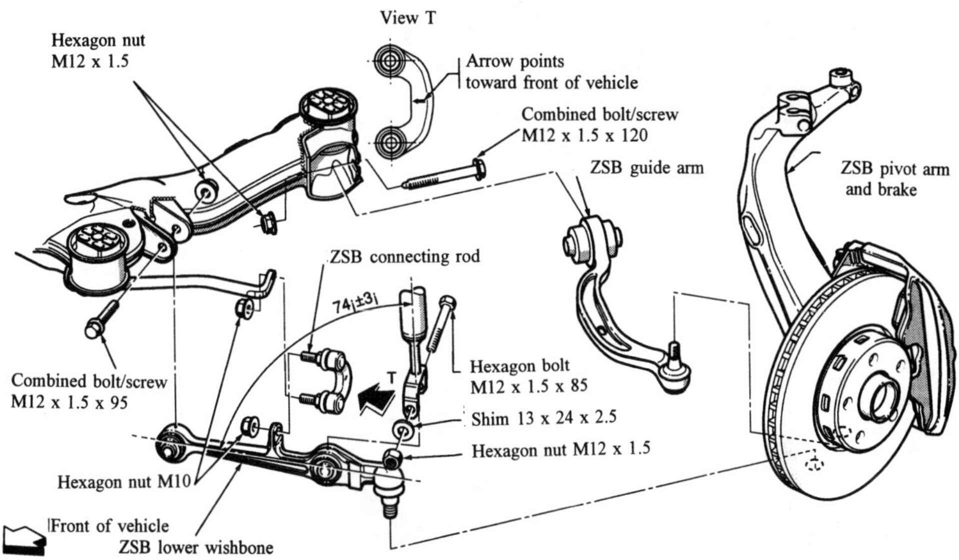

Figure 32 Connecting rod arrangement, front axle of Audi A8.

Heavily load-exposed connecting rods with elbows, Figure 32, are subjected to severe load under bending, despite the tensile-compressive force transmission. In this case, calculation by means of FE analysis is necessary, in order to make it possible to determine feasibility via simulation calculation and to develop the optimum lightweight construction.

The stress distributions in the front-axle stabilizing connecting rod (Figure 32) exhibit in critical areas Mises' stress intensity figures which cannot be met by short-glass-fibre-reinforced thermoplastics, see Figure 33. Plastics reinforced with directional long fibres, e.g. VE-GF60, can be regarded as an alternative to a forged aluminium component version. Both the weight and cost savings could in this case amount to approx. 20 to 25 %, signifying that this material substitution would amortize for high production-run numbers despite the cost and time intensive development input.

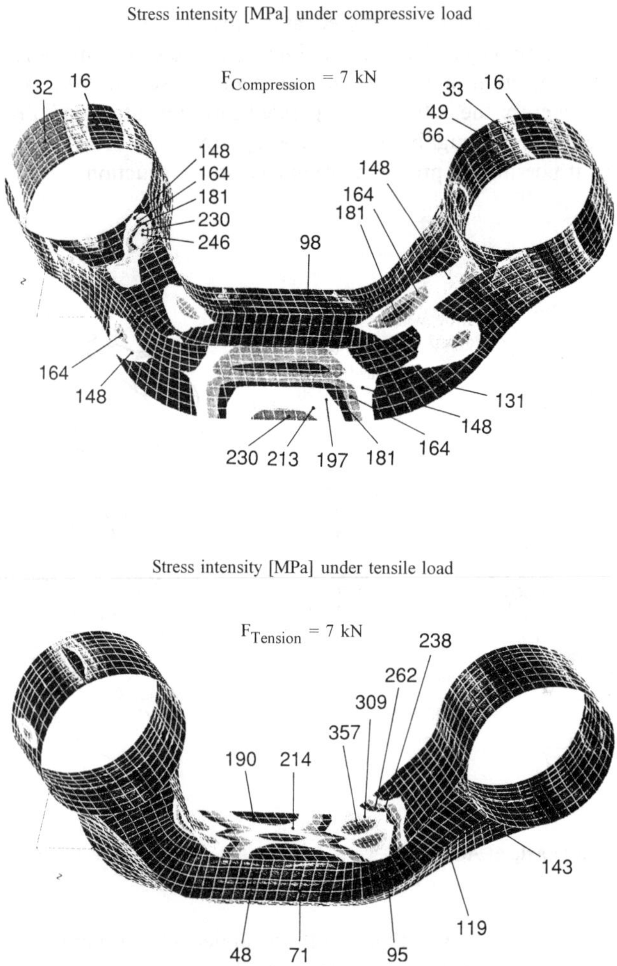

Figure 33 Stress distribution in dropped connecting rod.

Prospects

The pressure for fuel economy and emissions reduction, coupled with a demand for a high level of active and passive safety and high comfort in automotive engineering forces the development engineer to adopt functional solutions which convert the upward weight spiral to a downward one. This task can be solved only by using new, alternative materials and methods, in conjunction with completely new concepts and systematic integration of components and functions. The solutions and tentative solutions described demonstrate that both the fibre-reinforced thermosets and thermoplastics with their favourable energy/weight-specific stiffness factor ($\frac{\rho}{\sqrt[3]{E}}$) are eminently suitable [4], Figure 34. New opportunities, and at the same time challenges, are thus becoming apparent for all those active in this field, particularly in terms of the use of plastics in automotive engineering and with a view to materials recycling and holistic balancing.

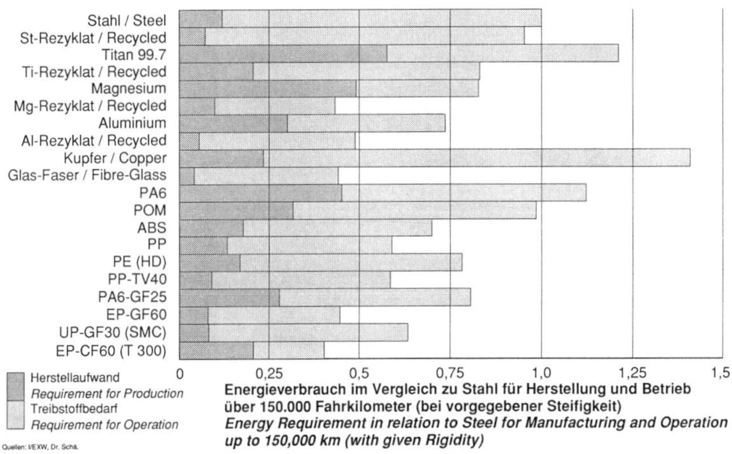

Figure 34 Energy balance for various materials as a function of weight-specific stiffness.

Symbols used

A	=	area content	e	=	peripheral fibre spacing
E	=	modulus of elasticity	f_e	=	natural frequency
I	=	moment of plane area	c	=	spring stiffness
F	=	force	m	=	mass
			b	=	acceleration
M_b	=	bending moment	ε	=	strain
			σ	=	stress

References

1 Haldenwanger, H.-G.: Entwicklung und Erprobung von Sitzen für Personenkraftwagen, ATZ 84 (1982) 9, pp. 437/ 445

2 Weichenrieder, A. and Haldenwanger, H.-G.: The Best Function for the Seat of a Passenger Car, SAE Technical Paper Series No. 850484, Int. Congress and Exposition, Detroit, Mich., 1985

3 Haldenwanger, H.-G., Leitner, P. and Mineif, P.: Plastics - a material for modern hybrid engine designs, Paper given at the Annual Conference Plastics in Automotive Engineering, Mannheim 1993, VDI Verlag ISBN 3-18-234170, pp. 19/107

4 Haldenwanger, H.-G.: Hochleistungs-Faserverbund-Werkstoffe im Automobilbau Entwicklung, Berechnung, Prüfung, Einsatz von Bauteilen, VDI Verlag ISBN 3-18-401251-4, pp 4/38, 79/120 and 121/146
5 Mundt, K.: FEM-Analyse einer Spurstange in Kunststoff-Metall-Verbundbauweise, Degree thesis, Technische Universität Dresden, 1993

Bumper paintwork using the example of the Mercedes-Benz C class

Ulrich Schatton, Lothar Jung and Peter Schüller

Design concept

In contrast to previous vehicle series, the C class was conceived in four equipment variants. Three versions were to be externally distinguished from one another by means of differing colour schemes for the bumpers (fenders) and the side moulding.

For styling reasons, the colour schemes for the C class bumpers resulted in three in some cases colour differentiated zones.

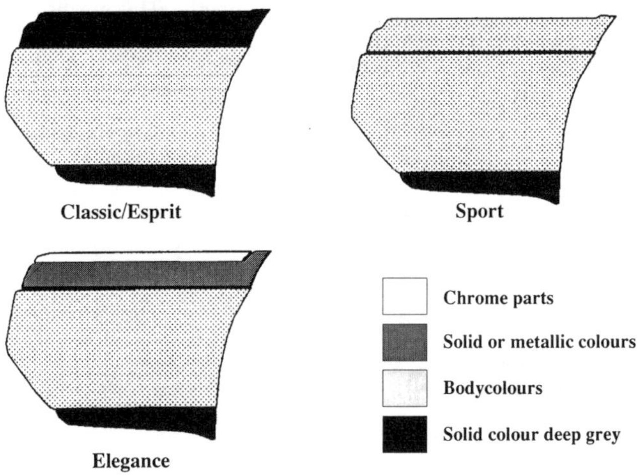

Figure 1 Colour zones of MB C class bumpers

Figure 2 C class; side view of the Elegance model

The upper section of the bumper continues after the wheel arches along the sides of the vehicle in the form of the side moulding attached. The colour of the lower zone is continued in the sill paint scheme. The width and colour of the central zone of the bumper is determined by the bodywork zone between the side moulding and the sill trim.

Depending on the particular version, this colour scheme for the bumper always results in a complex two or three-zone paint scheme.

Surround design

The painting costs for plastic attachment components for the vehicle exterior frequently amount to as much as approx. 50 % of the total component costs. For this reason, and in particular as a result of the number of zones to be painted in different finishes, great importance attaches to a surround design which facilitates painting. Most paint lines suitable for such components generally incorporate three or even four painting booths. This means that the primer, basis and clear coat are each applied separately in a paint booth. Where a fourth paint booth is installed, the base coat is applied in two booths, with a corresponding evaporation section installed between them.

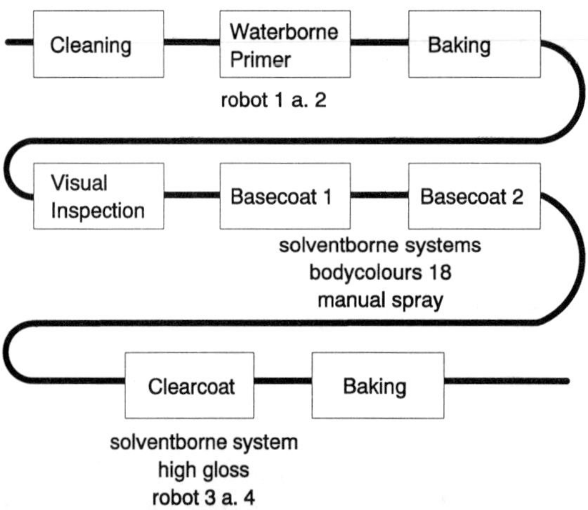

Figure 3 Four station system as used for the bumper surround

Where a component, therefore, is to be painted in two or even three different colours, and taking account of a broad range of colours, the question of the best method in terms of quality of finish and in terms of economy naturally arises. The "painting-oriented" design of the surround plays a decisive role.

The following alternatives are available:

a) a single-piece surround which would need to pass through the paint line described above two or three times, depending on its type. Masking operations would be necessary between the individual painting cycles.

b) a multi-component surround design for separate coloured components, in which each of the individual surround components only need pass once through the paint line. This would, however, give rise to a slightly greater assembly work input level.

Due to the bumper geometry and the configuration and size of the various colour zones, a two-part bumper surround for the front and rear component proved to be the best solution for all variants.

The following reasons indicated the suitability of the two-part surround design:

- easily manageable in the paint shop
- optimum utilization of existing paint lines
- simple repair painting
- masking requirements small and simple
- no separation edge problems
- low thermal exposure in the drying process
- simple logistics.

For the paint finish concept, it was important that the front and rear bumper surrounds were identically subdivided.

The best solution proved to be that of designing the central zone to be painted in the main vehicle finish as a separate component.

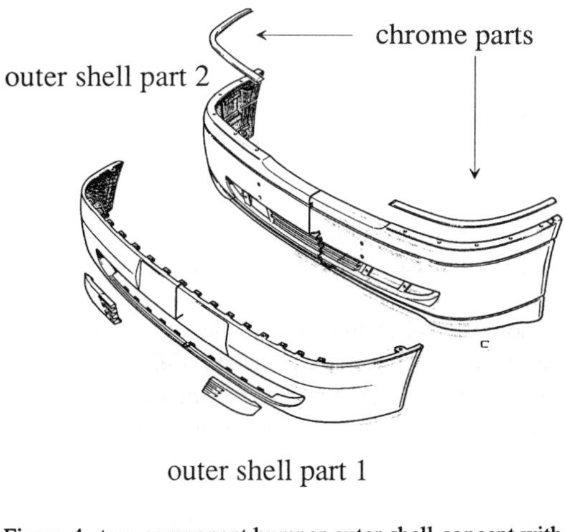

Figure 4 two-component bumper outer shell concept with integrated chrome strip

The sourcing concept provided for supply of the front and rear bumpers for all variants by two suppliers.

The planned daily requirement of 2000 bumpers plus the spares requirement meant that the two-piece design of the bumper surround was also justified on an economic analysis.

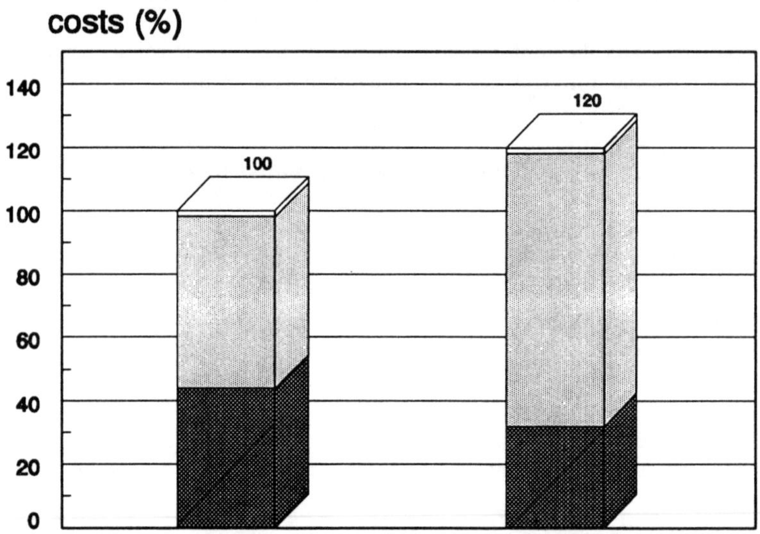

type		two part version	one part version
substrates	■	44	32
painting	▒	54	86
assembling	□	2	2

reference: total costs of two part version

Figure 5 Comparative cost analysis for various bumper types.
Example: Three-tone one- and two-piece bumper; Material: PC/PBT

This two-piece concept also signified a cost advantage for in-service repairs.

Additional subdivision of the component was not rational in the case of the surround, which has different paint finishes in the upper and lower component zones in the case of the Elegance and Sport variants. The width of the central zone means that painting of the upper and lower areas in different finishes is possible with an extremely simple masking system.

Requirements made on the painted component

A detailed discussion of all the requirement criteria which are applicable to such a bumper is beyond the scope of this article. This would be the function of the test, material, paint and dimensional specifications.

From a painting point of view, we placed particular emphasis on the following areas:

- The sections to be painted in the main vehicle colour should accord precisely with the bodyshell paint finish in terms of colour, gloss level, effect and texture.

- Any impairment of impact toughness caused by the paint system should be as slight as possible

- The high-gloss top-coat finish layers must be suitable for adequate polishing at room temperature

- Adequate intermediate-layer adhesion after over-painting with primer and/or top-coat finishes and in aged condition was vital, in order to permit repair paint jobs.

There were some difficulties in harmonizing the various requirements, due to their in some cases contradictory nature. Sandability with an overcuring-proof primer was initially not adequate, for instance. It was also initially possible to achieve a reasonable level of polishability only at the cost of inadequate impact toughness.

Paint concept

The paint concept depends essentially on seven principal factors. These are illustrated in Figure 6.

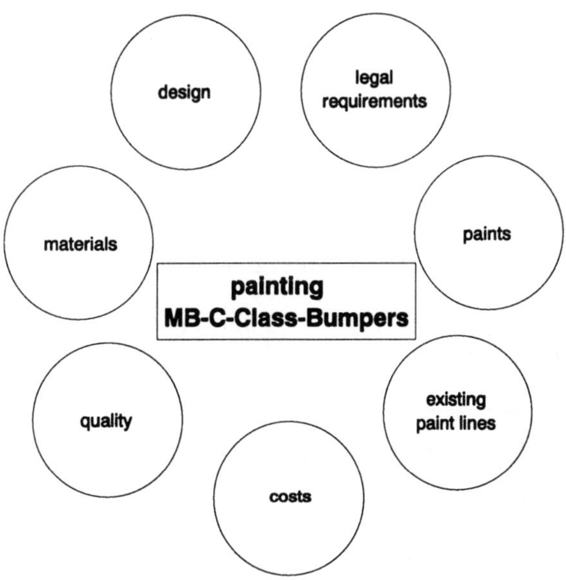

Figure 6 Paint-scheme concept

In this case, the styling targets for bumpers were extremely high. In some cases, they required thirty-three colours and three levels of gloss.

Of these seven principal factors, that of legal requirements will be dealt with in more detail.

The operation of paint shops is governed by an approval procedure in accordance with the fourth edition of the "Bundesimmissionsschutzgesetz" (BImSchG = Federal German Prevention of Pollution Act). This act at present exempts from the approval procedure paint shops in which less than 25 kg/h of pollutant organic materials are used. A simplified approval procedure is specified for paint shops, the emissions from which of organic substances are between 25 and 250 kg/h. Paint lines in which bumpers are painted on a commercial scale are within this range.

One of the main principles for the approval procedure is the Technical Directive on the Prevention of Air Pollution, referred to in German as the "TA Luft", for short.

The "TA Luft" on the one hand classifies organic substances into Substance Categories 1 to 3 and specifies the limits for mass concentration as a function of mass flows.

A permissible waste-air limit for organic solvents of max. 150 mg/m^3 results, on the basis of the "TA Luft". Areas in which manual work is performed are excepted from this provision.

"TA Luft" does not prescribe the way in which the limit is to be met.

The following are suitable methods of achieving these limits in paint shops:

a) System-related
Concentration of the organic solvent content, followed by waste-air cleaning, by means of thermal after-burning or absorption, for instance

and/or

b) Material-related
The use of aqueous (water-based) paint systems with a low organic solvent content.

This requirement is met in differing ways by various suppliers, as a result of the differing methods for achievement of the emission limits.

Many suppliers are reluctant to change from tried and proven paint materials. They perceive major uncertainty factors, such as the following, in the use of aqueous paint systems:

- Constancy of quality in the paint materials
- Modification uncertainty in existing paint shops
- Dependency on climatic factors
- Restricted durability
- Higher materials costs

These reservations are not shared by all suppliers of painted plastic components, however. The procurement, operating and maintenance costs for waste-air cleaning systems are high. For this reason, some suppliers use aqueous paint materials. Our experience is that interest in the use of aqueous-based paints and the seriousness of intentions of implementation of such systems are increasing among suppliers of painted plastic components. The paint systems approved by MB for car exterior plastics components are mainly three-layer structures, consisting of the primer, base coat and clear coat. This means that both purely solvent-based structures and hydro-systems, or combinations of both, must be developed in the current phase, depending on technical status.

The development status of aqueous paint systems varies from manufacturer to manufacturer. In addition, the utility and quality of the materials currently available on the market is differingly assessed by the various users. There are aqueous-based primers approved for nearly all the plastics substrates currently used for MB motor-vehicle exteriors.

In order to make allowance for the various methods of achieving adherence to the "TA Luft", we have had every exterior finish for our car ranges formulated in an aqueous-based grade by the individual paint manufacturers. Each new finish added is then formulated on both a solvent-based and a water-based system.

The objective of such a procedure is to be able to make a realization statement in both directions of development at a very early stage. Where this cannot be achieved, it will be necessary to implement a paint correction in conjunction with the design departments.

At short-term formulation on both a solvent and an aqueous basis makes it possible to set up painting concepts different for each supplier. This ensures that components with the same finish but supplied by different suppliers arrive at the vehicle manufacturer's in an acceptable quality.

This parallel development course should, in our opinion, be of only very limited duration, however.

Unfortunately, however, it is becoming apparent that this situation will probably continue longer than originally planned. The reasons can be found mainly in the desperate cost situation in which the motor-vehicle and supplier industries currently find themselves.

To use the example of the C class bumper, which is obtained from two suppliers, we have different paint structures, as a result of differing emission-reduction concepts, despite having one and the same plastics substrate. In the first case, both the mask and the surround are always painted entirely in solvent-based paints, irrespective of the bumper type.

In the second case, we are obliged to differentiate between the individual bumper components requiring painting and the bumper types, as a result of the paint lines installed. In order to ensure an adequate supply of components, painting is conducted in two paint shops using different paint materials. The reason for this is the difference in the operating permits possessed by these paint shops.

For this reason, the paint-scheme concept is as illustrated in Figure 7.

Component / Type	Mask	Trim	
		Upper section	Lower section
Classic and Esprit	Aqueous primer Three finishes Solvent-based base coat Vehicle finish Solvent-based clear coat (high-gloss)	Solvend-based single-coat top-coat deep-grey	
Elegance	As above	Aqueous primer Two finishes Solvent-based base coat Vehicle finish Solvent-based clear coat (silk-finish)	Aqueous primer deep-grey Solvent-based clear-coat, matt
Sport	As above	Aqueous primer Three finishes Solvent-based base coat Vehicle finish Solvent-based clear coat (high-gloss)	As above

Figure 7 Paint-scheme concept

This second case illustrates the large range of variants which can occur in the development of complex painted large-series production plastics components. Account has also been taken of the economic utilization levels of existing paint shops, adherence to official emission limits and the achievement of the quality required by the customer.

The paint concept coordinated and developed by the motor-vehicle manufacturer and the component and paint suppliers, is still not complete, however. Series supply is, on the present status, assured and coordinated with the MB plastics repair paints approved around the world for use by dealers and workshops.

Further planning for the C class bumper involves conversion of solvent-based base coats in contrasting and basic vehicle colours to an aqueous-based paint system.

The painting cycle and the requirements made on the paint systems

It was necessary to design, test and specify different supplier-specific painting sequences for the individual component variants and their differing colour schemes.

To take the example of one supplier, the painting sequence is as follows:

The single-colour components, such as the mask (all variants) or the dark grey surround (Classic and Esprit) are not the basic problem.

The mask, painted in the main vehicle finish, is painted, after cleaning, in a four-booth line, involving double application of the base coat, see Figure 3.

The Classic and Esprit surround was initially painted on a two-booth line with intermediate drying. A dark-grey one-coat paint now fulfils the requirements, however.

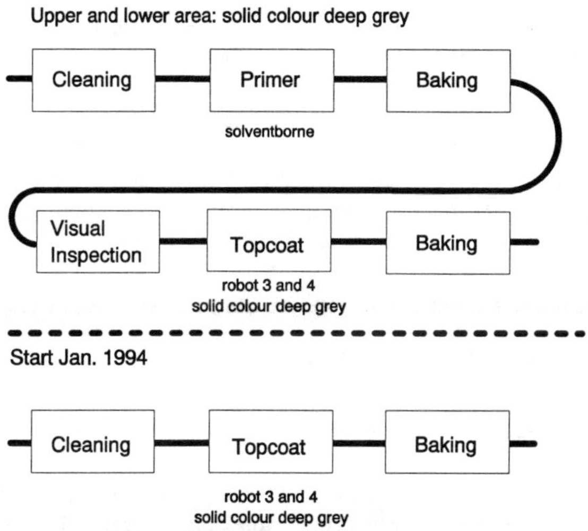

Figure 8 Painting process for the Classic/Esprit outer shell

The Elegance and Sport types, in some cases requiring two-tone painted surrounds, were new, however.
On the criteria of maximum cost-effectiveness and utilization of existing painting facilities, only one passage through a paint shop was specified for each component. This means, both in the primering phase and in the clear coat application phase, different materials must be applied simultaneously in one booth. This is accomplished in each case using two stations in each booth.

Due to the extremely broad colour range, the upper zone of the surround is painted with three colour-graduated aqueous-based primers. The lower zone is painted at the other station with a dark-grey aqueous-based primer. Rough masking must be applied to the upper zone before the base coat is applied, and then removed after application.

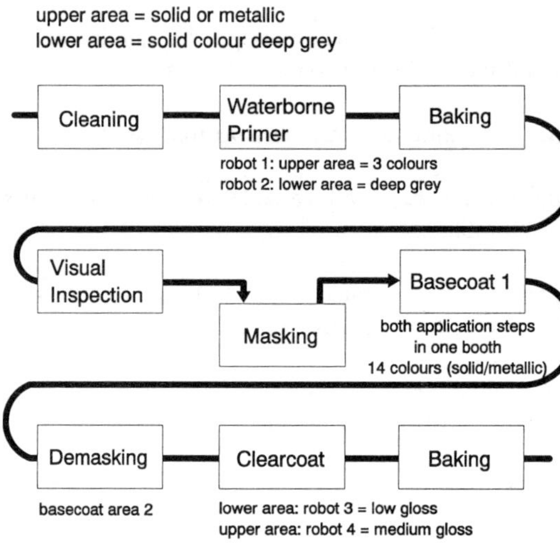

Figure 9 Painting process for the Elegance outer shell

Due to the differing gloss requirements for the upper and lower sections, the procedure in the clear coat booth is analogous to that in the primering booth. At the first station, the silk-finish (Elegance) or high-gloss clear coat (Sport) for the upper zone is applied. The matt clear coat for the lower section is constantly applied for both variants at the second station.

In addition to the normal requirements for the paint material, the process in this case added the following extra requirements:

- Masking-capability in the aqueous-based primer
- High application constancy in the base and clear coats, due to the need for frequent colour and gloss changes
- Broad application band width for the base coats in main vehicle colour with respect to colour constancy, as a result of the necessary divergent application systems for the mask and surround in the Sport variant
- Absorption of spray mist for the clear coats, the degree of gloss of which varies

Development and matching of paints

As a result of the differing colour schemes and degrees of gloss, and of two different processing sequences used by the supplier companies, the three different bumper versions gave rise to massive and complex effort in the development of the relevant paints.

No less than 65 different paints, including primer, base coats, clear coats and black top coats had to be readied for series use.

Properties	Primer	Base coat	Clear coat
Basis	Aqueous and solvent-based	Solvent-bearing	Two-component solvent-bearing
Colour	Pearl white Medium grey Dark grey	14 CC* 18 Vf* Colour Adjustment necessary for each supplier in the case of metallic finishes metallic: 20 24 uni: 3 6	
Gloss adjustment	matt	-- --	3 30 $\pm$ 5 E 1 53 $\pm$ 5 E 2 high-gloss 1
Total number	6	23 30 53	4
Special properties and Features	- Suitable aqueous primer with paintadhesion - Ageing and baking resistant - Hiding of raw component flaws - Support colour continuity - Aqueous primer as colouring element directly painted with silk-finish clear coat - Ageing resistant primer for spares	- High colour accuracy for both manual and robot painting - Good hiding power - Long storage life P.S. Continuous adaptation to valid colourcharts necessary	- Cold flexible - Polishable - Tar-resistant - Good stability - Wet-on-wet application of various degrees of gloss without masking - High degree-of-gloss constancy - Broad application pattern

CC* Contrasting colour
Vf* Vehicle finish

Figure 10. Overview of the paints in use

Two black finishes, with a single-coat and two-coat structure, must also be included in this list.

Technical paint-material tests

The first substrate alternatives (Phase 1) were tested for paintability in the period 1988 to 1990. The materials group was defined after approximately fourteen months. Laboratory tests were performed mainly on flat "tile"" materials during this phase.

Period	Materials group	Number of materials/paint systems studied		Objective	Result particularly in terms of suitability for aqueous primer and use of aqueous base-coat
1988/89/90	PP/EPDM	3	30	Paintability statement Approximate estimate of necessary development input, definition of the paint systems to be used, determination of suitable pretreatment methods	Previous aqueous primers were suitable at the time of the study only for these substrate groups. Development of suitable aqueousprimers and aqueous base-coats was started.
1990/91/92	PBT/PC	approx. 80	approx 120	Paintability statement Approximate estimate of necessary development input, definition of the paint systems to be used, determination of suitable pretreatment methods	Suitable aqueous primers also had to be systematically developed for this materials group

Figure 11. Technical paint-material tests

Period	Materials group	Number of materials/paint systems studied		Objective	Result particularly in terms of suitability for aqueous primer and use of aqueous base-coat
1990/91/92	PBT/PC	4	approx 400	Definition of - paint systems - plant parameters - paint application methods	- Final selection of substrate - Start of system and prototype tests - Initial clearance

Figure 12 Planning of the painting process sequence and system tests

Planning of the painting process sequence and system tests

In this second phase, the planned process sequences were simulated jointly with supplier companies in existing equipment systems, using W124 and W201 series bumper components. The knowledge obtained increasingly resulted in the sequences and paint materials actually adopted.

Conclusion

The analytical study on the bumper trim performed in the autumn of 1989 by MB resulted in the two-piece bumper version.

This component division offered advantages on all of the following criteria:

- quality
- cost-effectiveness and
- technical production

The involvement at an early stage of the potential suppliers, particularly those in the paint sector, was of special benefit in this context.

The large number of colours to be offered was a manageable problem. At least half of the colours to be used had already been introduced into the bumper-suppliers' ranges in connection with the bumper of the previous model.

The MB vehicle finishes were new for the bumpers but were familiar to the suppliers as a result of other plastics components already in use.

The painting tests performed in advance of the final selection of the substrate on the basis of cooperation between the paint industry and the substrate manufacturer proved after definitive selection of the material to be a significant advantage in terms of rough advance delineation of potential paint systems. It thus became possible to shorten the time input for development of the paints, which had inevitably been lengthened as a result of the suppliers' different paint concepts.

Due to the fact that experienced bumper suppliers with tried and proven painting plant familiar to both the paint supplier and the component user were commissioned with this project there were no critical or unmanageable phases in the implementation program. Many years of cooperation, knowledge of the relevant development trends, and the methodical procedure adopted by all participants were the decisive factors in the trouble-free achievement of the objectives set.

The decision to use solvent-based base coats has proven, on a retrospective analysis, to be correct for the assurance of reliable series-production start-up.

Once the aqueous-based primers and solvent-based base coats introduced have successfully passed their "probationary" period, work can be started on the achievement of utilization of an aqueous-based base coat.

The extremely high level of continuous series-production liaison input on the part of all those involved for such complex painted components should not be underestimated, however.

From a technical paint point of view, it is desirable that the period between fine tool coordination and series-production supply should be greater, or should be better exploitable, particularly in the case of general occurrence of difficulties.

Prospects

The objective must be to achieve new targets, now that series start-up has been overcome.

The principal objective will be cost reductions.

Potential provisions with their derivative functions and period of implementation are as follows:

Potential provisions are	functions derived from these	Period
- Reduction of the number of coats	One- and two-coat systems	1993/94
- Improvement of application efficiency	HVLP*, Esta, robots	Ongoing
- Reduction of emissions of organic solvents	Single-coat aqueous black system Development and use of aqueous base-coats	1994/95
	Development of aqueous clear coats	1995/96

*HVLP = High Volume Low Pressure

Low-pressure processes for decorative interior fittings

P. Strasser and B. Woite

The interior trim of vehicles has always been the domain of organic materials.

From the dashboard with its crafted wood up to and including ultra-luxurious leather seat trims, the interior has always reflected the changing tastes of succeeding decades. In the course of time, however, the "mobile living room" has increasingly been required to fulfil demands which were now beyond the capabilities of the natural materials. The temperatures to which the materials in a vehicle interior are exposed, for instance, range from an arctic -30 °C to a tropical 130 °C, thanks caused by the large, aerodynamically optimized glazing of modern-day passenger compartments.

The ascendancy of the synthetic materials, with their broad and versatile property profiles, began to accelerate in the nineteen-sixties, with the use of artificial leathers, plastic film and sheet, and textiles. Since the interior trim of present-day vehicles already consists practically exclusively of plastics, no further growth in volume may be anticipated in this field. The prevalent factor now is replacement of individual materials, on the basis of a large number of criteria. In some cases, highly diverse results occur for each individual vehicle manufacturer and even model series, due to the complexity of the requirements and differing priority systems.

Interior styling is dominated in principle by visual appeal, quality and the harmonious combination of selected surfacing materials. The motorist tolerates and even accepts plastics, but he definitely does not want a "plastic" look. The interior trim has a significant ranking in the factors influencing the decision to purchase made by a future motorist (and not only for the female gender). For this reason, the selection criteria should be examined not least of all for their relevance to the customer (see Table 1).

The first priority with the customer is the price/benefit ratio, efficiency also being moderated emotionally by ambience and comfort in the interior design.

Here, the selection of surfacing materials for structure, colours and quality of finish plays a dominant role. Meeting of safety standards, positive flame resistance and the recyclability of materials are all properties which are presupposed and therefore have no immediate influence on the decision to purchase.

The selection of materials is in principle of no significance to the customer provided they do not clash with the above-mentioned criteria. This signifies even more opportunities for motor-vehicle manufacturers to improve the price/benefit ratio without impairing customer-relevant properties in the field of materials and the technologies used.

The low-pressure processes are capable of making a significant contribution to the more rational and cost-effective manufacture of trim components in vehicle interiors.

Table 1 Customer-relevant characteristics of interior trim

Characteristics / Customer relevance	**Selection criteria**	**Property profile price / benefit ration**
Primary selection criteria	Price / Visuals / Haptic	Selection of surfacing materials for structure, gloss, colour, finish; harmonization of various surfacing materials, texture, functionality (e.g. pockets, etc.), on-road noise (grinding noises)
Secondary selection criteria	Safety / Acoustics	Fulfillment of safety standards, lightwight potentials, noises suppression and attenuation, insusceptibility and robustness, emissions (odour nuisance and fogging), recyclability
No quantifiable customer relevances	Materials / Technologies	Technologies, workability, backing and tactile materials, easy assembly and dismantling

Interior trim zones

The interior of a vehicle is composed in basic-model vehicles of cheap and single-layer trim components. These generally take the form of grained injection-mouldings not capable of fulfilling high-level demands on their visual and tactile appeal, acoustics and safety properties.

Principal attention in this study will be directed to the multi-layer materials combinations such as are used in medium-range and top-end vehicles. Such interior trim components have the following structure:

- Surfacing material (sheet, skin, textile)
- (Possibly) tactile-enhancing intermediate layer (foam or fleece)
- Backing materials

Classical applications of this type in the vehicle interior are

- door and rear quarter panels
- A, B and C pillars
- rear parcel shelves
- foundation board for front seats
- trim panels under dashcover
- trim in the boot

All these selected applications have in common that they are generally "flat" components with the least possible component depth. For this reason, applications such as dashboards, cockpits and consoles are heavily dependent on styling for their integration into the overall scheme.

The requirements made on the interior trim

Interior trim components are required to pass a large number of physical and chemical tests in order that they fulfil the customers' quality demands under all driving conditions and situations throughout the life of the vehicle. The most important of these are (taking the example of a door trim panel):

- Ageing tests using simulated sunlight and climates
- Behaviour under storage in water
- Short-term strength tests on adhesion and attachments
- Fatigue tests
- Acoustics tests
- Tightness tests (exclusion of drafts and moisture)
- Functional tests
- Emissions
- Combustibility
- Side-on impact behaviour
- Recyclability

The principle of the low-pressure processes

The increase in laminar structure doubtless achieves enhanced quality in the interior trim but this improvement is accompanied by higher costs (see Figure 1).

The basic principle of the low-pressure processes is that of producing cost-effectively and rationally, wherever possible in a single operation and in a single tool, components which, with a multi-layer structure, are capable of fulfilling all the quality criteria.

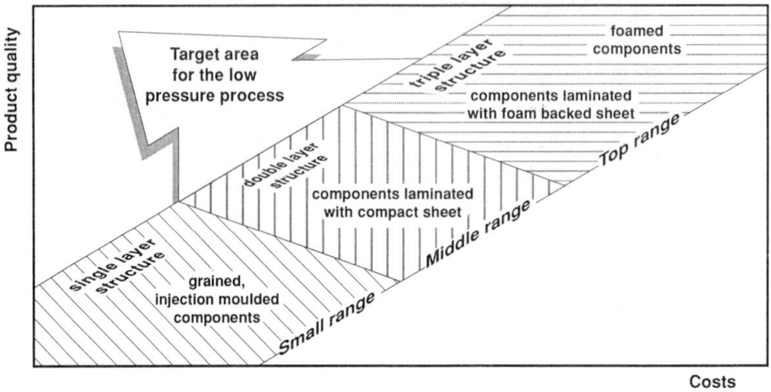

Figure 1 Qualitative structure of the interior trim as a function of cost

The following two factors are necessary for the most nondestructive possible treatment of the surface materials to be filled:

- low temperatures (in order not to thermally damage surfacing and foam materials) and
- low pressures (in order to preserve surface textures and haptic properties).

Figure 2 shows a summary of the possible processes, which are discussed in more detail below.

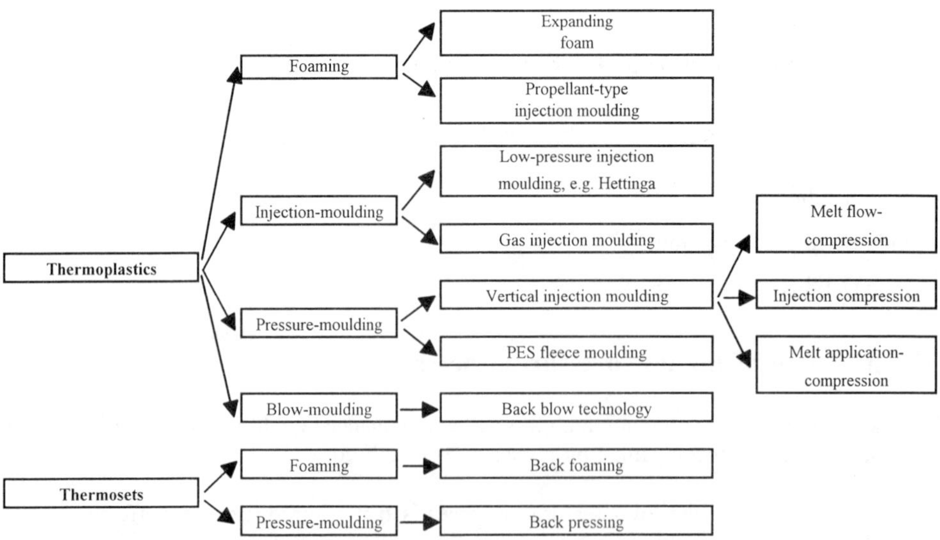

Figure 2 Overview of the low-pressure processes

In order be able to produce trim components cost-effectively in a single tool, it is necessary to examine the laminating manufacturing process nowadays widely used for its rationalization potentials in more detail.

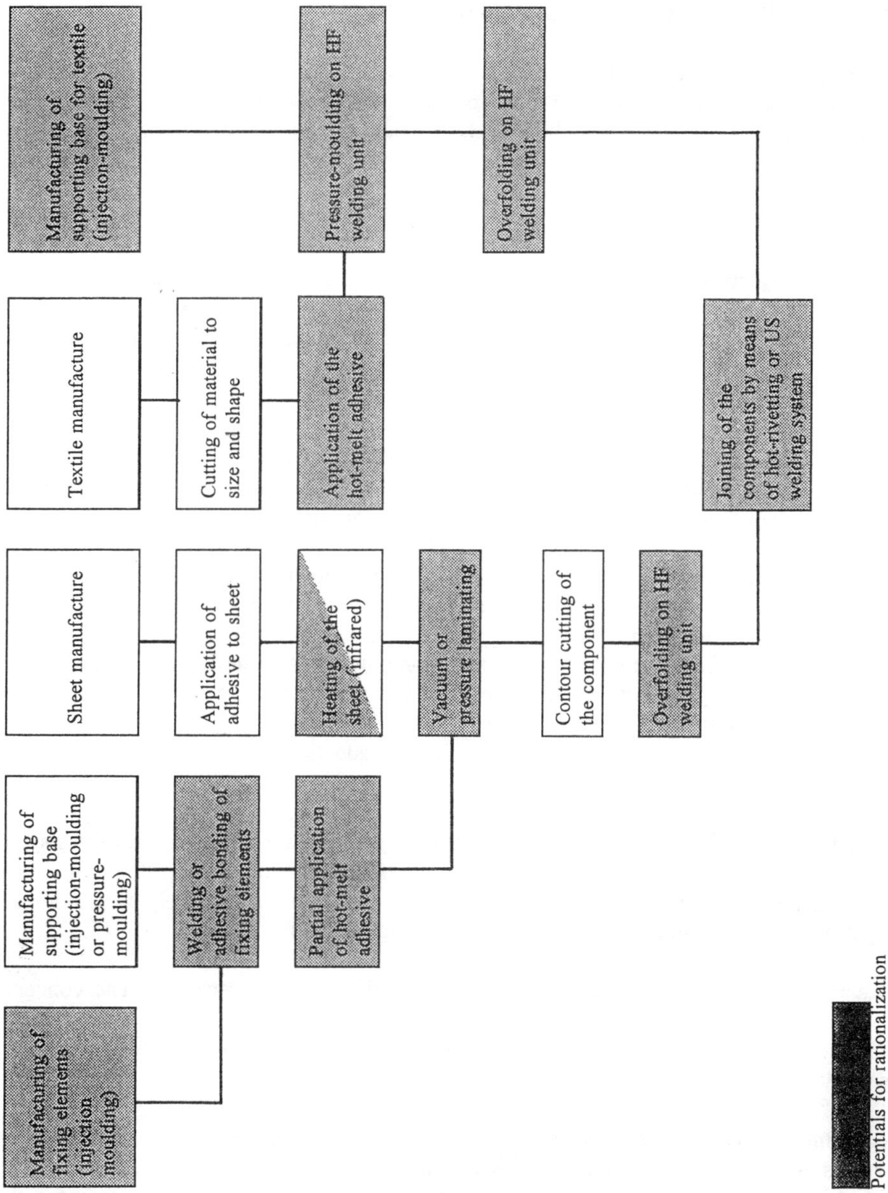

The following is intended to establish which of the cost potentials illustrated can be achieved using the various low-pressure processes.

Summary of the low-pressure processes

Conventional injection-moulding for processing of thermoplastics is the basis for the low-pressure processes. The objective in their development was the reduction by process-engineering means of the excessively high internal mould pressures from over 300 bar down to a pressure level which would permit back-filling of standard sheets, foil and textile structures.

Low-pressure injection moulding

If the filling pressure used in filling of the cavity can be minimized by means of process optimization, one enters a pressure range well below that of the standard injection-moulding process.

Hettinga was the first machinery manufacturer to recognize that low-pressure injection-moulding in combination with in-mould decoration of decorative materials offered a considerable potential for development, particularly for the motor-vehicle industry.

The mechanical technology was derived essentially from propellant-type injection moulding machinery [1]. A special screw with a mixing section is intended to achieve an extremely high melt homogeneity at the lowest possible melt load exposure. In addition, the melt temperature is 20 to 40 °C lower than in conventional injection-moulding. A modified machine concept in which up to six clamping units can be served in sequence by a rotating injection unit makes it possible for the plastification process to take place in principle only slightly before the subsequent injection operation. This achieves a further reduction in the thermal exposure of the moulding material.

The machine has a high-capability control system incorporating ten speed and pressure stages for the plastification and injection sequences plus the interlinking of screw path and locking pressure. The latter feature permits compression injection moulding using shear edge moulds, making it possible to further reduce the pressure acting on the textile.

Hettinga machines have been widely used in the textiles-using industry [1]. The concept failed, however, in the long term to become established, and several well-known machinery manufacturers nowadays supply equivalent or even superior solutions for in-mould decoration of decorative materials.

Further-refined processes are nowadays already used for numerous vehicle interior-trim components with a textile surface on a series basis - in the A and C pillar trims of the BMW 3 Series, for instance (see Figure 3). Further examples include trim components with plush surfaces in the footwell. These are in nearly all cases applications involving minor components with only slightly projecting surfaces and small levels of deformation.

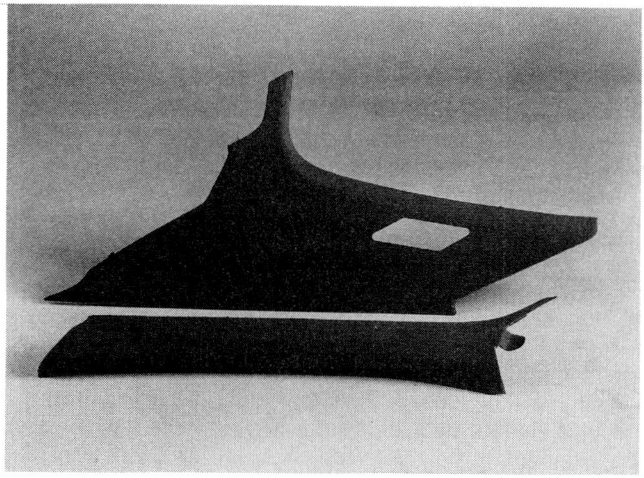

Figure 3 A and C pillar trims using the in-mould decoration method

The process sequence is, in principle, similar to that for the conventional injection-moulding process (see Figure 4). Before the tool closes, cut textile blanks either stacked or fed from a magazine are positioned by means of needle grabs between the two halves of the mould and fixed there. This can be accomplished by means of needles, hooks, frames and other similar systems.

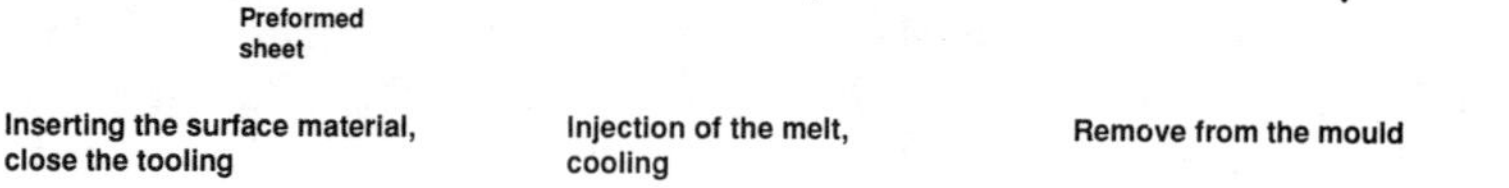

Figure 4 Process sequence in the in-mould decoration method

In order to avoid damage to the decorative material, the hot-melt pressure at injection should not exceed 100 bar. The decisive factor in achieving a low filling pressure is selection of the correct rate of injection. Due to the structure viscosity of plastic melts and shearing-induced heating, the necessary melt filling pressure does not rise linearly with volumetric flow. The increasing pressure at extremely low rates of injection can be explained in terms of cooling of the heat against the wall of the mould and the increase in the thickness of the solidified boundary layer. For the in-mould decorating process, it is particularly advantageous if the hot-melt flows along the interface with the insulating textile and the growth of the solidified boundary layer is thus retarded. This is the reason that the melt filling pressure for the in-mould decorating process is in practice lower, with an identical rate of injection and wall thickness, than is in theory necessary, and the reason why the minimum pressure is shifted towards even longer injection times. Melt pressure in the tool can be most effectively reduced if compression and post-injection pressure are omitted after filling pressure has been optimized. In after pressureless injection-moulding, there is no compensation for volumetric shrinkage during the cooling phase. In addition, a reduction in post-injection pressure has a beneficial effect on the stress and orientation state of the moulded components [2].

A reduction of this pressure can also be obtained by means of controlled breathing of the mould or the use of the compression injection moulding technique. In compression injection moulding, the cavity is initially only partially filled. Forming of the moulding is then accomplished by the closure motion of the moving half of the mould.

The mould is a further important component in the in-mould decoration technique. There are certain basic differences vis-à-vis conventional injection-moulding tools. Injectors, for instance, must be integrated into the nozzle-side of the mould, because of the decorative element.

On the basis of the present-day standard of modern injection-moulding machines, various machinery manufacturers include in their ranges expanded control technologies specifically for this application. The spectrum extends from minimization of filling pressure via systematic optimization of rate of injection, melt temperatures and mould temperatures, on through to cascade and embossing control systems. Low locking forces relative to component geometry are necessary, due to the low injection pressures. Special machines with extra large mould carrier plates are available to nonetheless permit full exploitation of the potential closing force of a particular machine.

In-mould decoration of decorative sheet and foil, which is used, for instance, for decorative trim strips in vehicle interiors, can be regarded as a variant of the process described, although not necessarily in the low-pressure range. The term "In-mould decoration" (IMD) stands for injection-moulding and hot embossing in a single operation. IMD sheet (manufactured by Kurz, for instance) is fed into the injection-moulding tool. During moulding, the lacquer coating separates from the polyester backing film under the influence of the pressure and temperature of the moulding compound and combines accurately with the moulded substrate. Three dimensional components can be produced up to a depth of 5 to 10 mm. The so-called Sublex sheets produced by GE Polymertrend, which are thermo-formable, offer even more potentials in this respect. The deep-drawn sheet is placed in the injection-moulding tool and in-mould decorated. Both processes are capable of applying any photographically recordable decorative elements to the component.

Gas injection moulding (GIP)

Gas injection moulding technology is also a specialized variant of conventional injection-moulding among the low-pressure processes, the principal emphasis in the process's development being in the manufacture of rationally priced components with a high rigidity. High component rigidities can be achieved either by means of high moduli of elasticity (generally more expensive materials incorporating reinforcing agents) or, at practically no extra cost, via high moments of inertia (hollow-cavity structures).

In the gas injection moulding process, the plastic core is penetrated towards the end of the injection operation by an inert gas (nitrogen). This produces a cavity, by means of which mass shrinkage is balanced out using gas pressure. Low-stress and therefore low-distortion components are produced, thanks to the constant low pressure level of the gas. In contrast to compact-method injection moulding, the gas injection moulding process is capable of reinforcing flat components with high-rigidity hollow-cavity structures.

Sunk spots may occur, however, around the gas ducts in in-mould decoration of foam sheets and textiles. The insulating action of surfacing materials produces large differences in the rates of cooling of the melt around the duct and the adjacent thin-walled component sectors. For this reason, gas internal pressure ducts should not be positioned in smooth, low curvature areas, but instead only along visible edges and boundary areas, in order to ensure the obtainment of high-quality components.

The pressure level produced in the cavity depends above all on the number, location and magnitude of the gas ducts relative to the homogeneous component-material areas. An A pillar trim (with edge ducts) can, for instance, be produced at a lower pressure level and therefore with more decorative-element safety than a large door-trim with edge reinforcement. For this reason, this process would appear to be more suitable for the manufacture of smaller interior trim components bearing decorative elements.

Propellant-type injection moulding

The propellant-type injection moulding process is also a low-pressure process, since an internal pressure of a maximum well below 100 bar is generally reached in the mould [3]. A moulding compound which includes propellant gas is injected into the cavity, generally by means of a central injector, the propellant gas aerates the hot-melt and thus completely fills the tool cavity. Since injection must take place at very high speed, thermal and/or physical damage to the surface materials may occur around the injector. Long flow paths necessitate high component thicknesses in order to avoid excessive shear loads acting on the decorative elements.

The disincentives for the use of propellant-type injection moulding technology as a production method for interior trims are, above all, high component costs (as a result of long cycle times and the need for finishing), low potential for lightweight components (large wall thicknesses) and the detriment caused to the environment (propellant agents).

For this reason, this process has failed to achieve significance for production of vehicle interior trim components.

Expanding thermoplastic foams (particle-filled foams)

Particle-filled foams on a polystyrene basis have now been familiar for many years under the product name "Styropor" (in Germany; known as expanded polystyrene foam in the U.K.).

Despite their great potential for lightweight designs, these materials have become established in automotive engineering only in special applications (such as children's safety seats, for instance). The reasons for this can be found primarily in their poor physical properties (strength and toughness) and in their inadequate dimensional stability under heat. Only particle-filled foams based on polyethylene and polypropylene exhibit property profiles which permit their use in motor-vehicles. Components made of these materials are notable for an extremely low weight, high impact absorption, good breaking strength and adequate temperature of deformation under flexural load.

Expanding PP random types, which have the most interesting property profile, may be considered for vehicle interiors. Single-type composite structures can be created in combination with TPO sheet. Preformed TPO sheet can, for instance, be placed in one half of the mould with PP foam backing and vacuum-attached (see Figure 5). The two halves of the mould are then closed. The foam particles are then drawn in and conveyed into the tool cavity by the filling apparatus, which consists essentially of a storage tank and the pneumatic filling injectors.

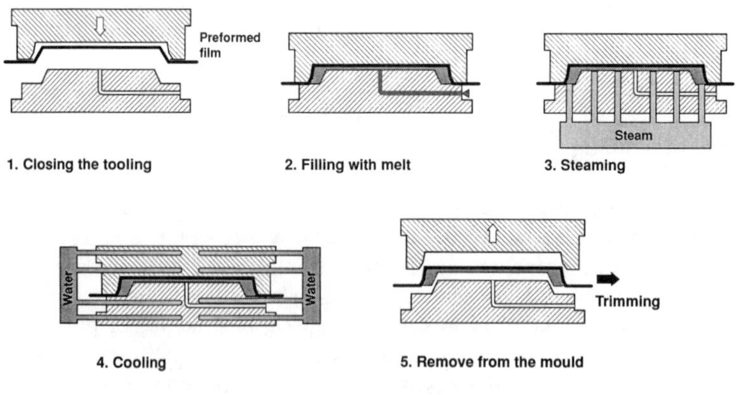

Figure 5 Back foaming using EPP particle-filled foam

Filling of the mould is followed by welding of the foam particles with one another and with the PP foam of the TPO foam sheet by the thermal energy of the steam, this being applied only from the filling side.

The mould and the moulding are at a temperature of 120 to 150 °C at completion of welding. The component must now be cooled until it is geometrically stable. EPP foams require cooling down to around 50 to 60 °C.

The moulding is ejected from the mould by compressed air, assisted generally by ejectors, and transferred to a removal system if one is installed.

Automatic moulders are machines for the intermittent production of mouldings from particle-filled foams. One differentiates between vertical and horizontal types, depending on the opening and closing motion of the mould sections. With the exception of a few special cases, automatic moulders are primarily of the horizontal type. Their essential components are:

- the clamping unit and steam chambers
- the filling system
- piping and valves for air, steam and cooling-water
- the ejector mechanism and
- the program sequence control system.

The use of particle-filled foam is particularly suitable for energy-absorbing components (e.g. side-on collision padding, knee-protection pads, etc.). The combination of EPP backing materials with TPO foam sheet is rational for door and side-panel trims, in which the tactile element is achieved solely by the foam sheet.

Due to the materials positive property profile, the applications for particle-filled foams on a polyolefin basis will with certainty increase in the future.

The vertical injection moulding (VIM)

As already mentioned, the VIM process can be used to achieve a significant reduction in process pressure. The magnitude and duration of the embossing gap opening is, however, severely restricted by the potential flowing off of the hot-melt if a horizontal closure motion is used. For this reason, the VIM-process is notable as a refinement with a vertical closure motion, injection taking place, here, too, into the open tool.

Due to its low pressure level, VIM permits the use of decorative materials with a tactile foam which produces so-called "soft touch". Since the decorative elements are generally secured to the upper section of the tool by means of clamping frames, hooks, vacuum or a similar method, a significant advantage can be found in the fact that contact between the decorative material and the hot backing material only occurs at a relatively late point after completion of insertion of compound into the tool, thus avoiding thermal damage to the tactile foam.

Since compound is charged into the open mould, internal mould pressure which, thanks to its low level, permits nondestructive processing of the decorative elements, can build up only once the mould begins to close. Plastification is always performed by an injection unit in which the plastified compound is metered into the mould either as a round strand or as a ribbon through a flat sheet die or a hot-runner.

For recyclability reasons, material structure is increasingly being selected in such a way that the material combination of decorative sheet, tactile foam and backing material consists of a single plastics "family". The backing material in in-mould embossing may, for instance, consist of an impact-resistant talcum-reinforced polypropylene while the decorative sheet is made up of a thermoplastic polyolefin onto which a polypropylene foam is laminated.

One differentiates in vertical injection moulding between the <u>Melt Flow Process</u>, the <u>Melt Application Compression</u> and the <u>Preform Injection Compression</u> [4].

Vertical injection moulding using the melt flow process

This processing principle was developed in Japan and is also known as the SP-MOLD or Sumitomo process. The hot-melt is charged into the horizontal plane with the mould open into the lower (fixed) half of the mould via a hot-runner system (see Figure 6). Rate of injection must be selected in such a way that a jet is avoided. Large mouldings require several gate points in order to achieve complete filling of the mould. In addition, needle-type shut-off nozzles in conjunction with a cascade configuration permit uniform moulding filling even with complex mould geometries.

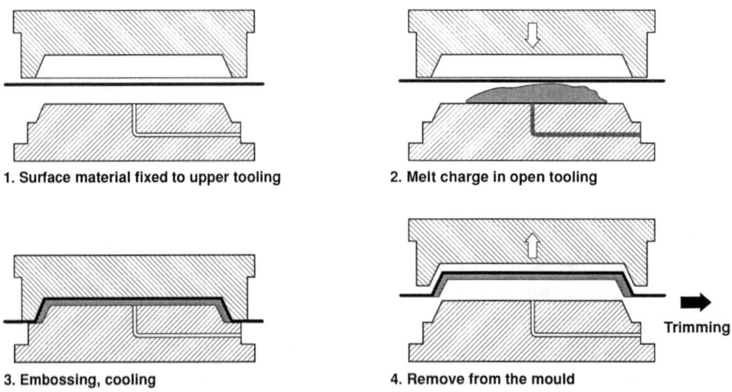

Figure 6 VIM using the melt flow process

The cakes of compound should not be allowed to come into contact with the decorative element prior to embossing, in order to avoid damage as a result of excessively long exposure to temperature. For this reason, the mould gap should be selected large (approx. 30 to 50 mm) and the sprue gates balanced out in such a way that impairments as a result of melt seams can be avoided even without a cascade configuration wherever possible.
During the actual embossing operation, the flowing hot-melt exerts a shearing action on the decorative layer. The possibility of wrinkling, particularly at seams, can be avoided by retarding the closure and using a slow closure speed.

Vertical injection moulding using the melt application compression process

The melt application compression process, also known as the hot-melt deposition process, was developed to eliminate the melt seam problem in principle. A further advantage of the strand piling process is the fact that the moulds can be constructed extremely simple, i.e., without any complex hot-runner system. Plastification of the backing material occurs in an extruder or, to achieve more precise proportions, in an injection unit. Application of hot-melt into the lower mould section is accomplished by means of the reciprocating motion of the plasticizer unit in the axis of the screw. After deposition of the strand of compound, which may take the form of a round strand or a ribbon, depending on die design, the mould is closed and the embossing force is applied (see Figure 7).

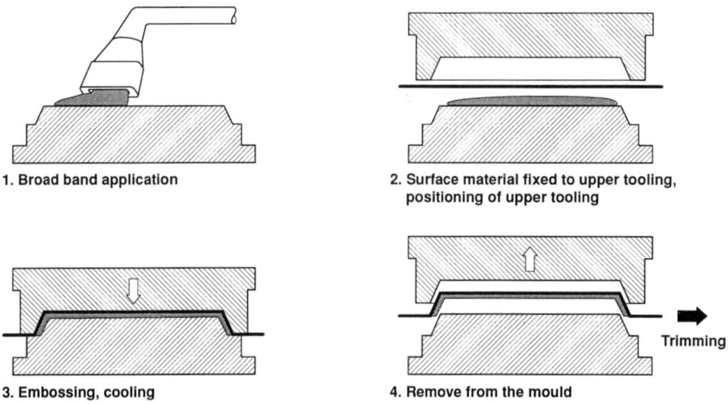

Figure 7 VIM using the melt application compression process

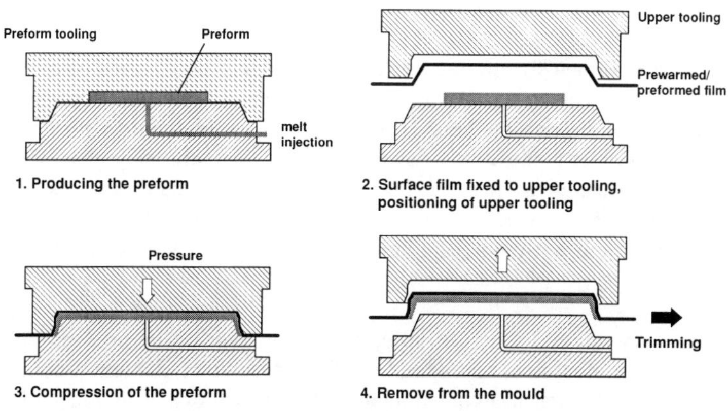

Figure 8 VIM using the preform injection compression method

In the case of large components with fluctuating widths (such as a rear quarter panel with a wheel-housing recess, for instance), width control of the strand of compound is of great significance. There are no distance-dependent width and thickness control systems for the flat sheet dies currently available on the market, which have generally been developed for extrusion purposes. There is still a need for development in this area; allowance must be made for geometrical recovery capacity and broadening of the strand. A certain limited potential for adaptation of the strand to component width is provided solely by means of a distance-dependent injection-speed plot or by means of a lateral movement of the injection unit. For mouldings with small dimensions, it is sufficient for the hot-melt to be deposited in the form of a strand from a round die. The rate of deposition can be adjusted to the compound requirement.

One problem with this process is the repeatability of the geometry and position of the strand of compound.

Vertical injection moulding using the preform injection compression

The preform method permits more precise positioning and proportioning of the hot-melt in the mould; it can be subdivided into two process stages, i.e., the making of the so-called preform and the actual in-mould embossing process (see Figure 8). The making of the preform is accomplished in an injection-moulding tool which consists of the original lower mould half (backing material side) for in-mould embossing and a special upper mould section. The upper mould takes the form of a simple preform cavity into which injection takes place using the standard injection-moulding method, i.e., injection into a closed mould. A solidified boundary layer forms as a result of cooling of the hot-melt against the surface of the mould.

In order to achieve a plastic core in the preform, the wall thickness of this preform is increased and cooling time is reduced to a minimum. The initial surface area of the preform should be 30 to 50 % smaller than that of the finished component surface, in order to achieve a greater wall thickness for the same volume [4]. After the cooling period, the mould is opened and the original mould section with the fixed decorative element moved into its working position; the moulding remains on the lower section of the mould.

The decorative sheet may be preformed for more complicated moulding geometries or only preheated in the case of simpler geometries. During the subsequent in-mould embossing process, the plastic core penetrates through the boundary layer and fills the mould with hot-melt. Compound temperature is thus significantly lower in in-mould embossing than the temperature necessary for plastification of the backing material. This circumstance is of assistance in the nondestructive processing of heat-sensitive foam sheets. The typical penetration of the fabric through the hot-melt (capillary effect) typical in textiles with no barrier fleece is also reduced.

In this process, the visible marks caused by the solidified boundary layer of the preform on the decorative element side of the moulding, and adhesion of the decorative element around the preformed injection moulding (which has cooled against the surface of the mould) still cause great problems in this process.

The advantages of the individual VIM processes:

Melt Flow Compression	Melt Application Compression	Peform Injection Compression
Precise proportioning	No hot-runner, no preform mould	Precise proportioning
Short flow parts		No seams
	No seams	

The disadvantages of the individual in-mould embossing processes:

Melt Flow Compression	Melt Application Compression	Peform Injection Compression
Complex hot-runner mould	Width of strand non-variable	Extra preform mould
Allowance must be made for seams	Repeatable positioning of the hot-melt not possible	Marks as a result of solidified boundary layer

This list indicates the most important qualitative and economic advantages and disadvantages of the individual VIM processes. Since this technology is still extremely recent and only few series applications of it have occurred, at least in the German-speaking countries, final assessment at this point would most certainly be premature. The preform injection compression must, at present, be rejected as a result, above all, of the visible markings caused by the solidified preform boundary layer. The melt application compression process will be used primarily for long narrow components. Here, the development of flat sheet dies with thickness and width controls and a precise strand cutting system could open up new potentials for large area components for which, up to now, the melt flow method was primarily suitable.

A detailed examination of economy assessed against the other (low-pressure) processes will follow separately below. It should, above all, be noted that in all VIM processes employing turntable or slide-table concepts, compound charging and the embossing operation are performed separately, in order to achieve economically viable cycle times.

Pressure thermo-forming of fleeces (fleece moulding)

The objective of this process is mainly the production of interior trim components with a textile surface in a single-type and therefore recycling-friendly material structure[1]. Since

[1] Jägerhuber, G.: Thermoformpressen von Nadel- und Thermofusionsvliesen zu Innenverkleidungsteilen in Sandwichbauweise mit Träger- und Haptikteil aus sortenreinem Polyester. Thesis at Technische Universität Coburg, 1993.

surface textiles consisting of polypropylene fibres are not yet capable of meeting the requirements for interior materials, the process development work has been based on a single-type polyester fibre structure.

For this purpose, the polyester decorative material and the assembled polyester fleece stacks, both possessing differing properties, should be positioned in a mould and processed in such a way that a three-layer structure, consisting of the surface, the tactile zone and the backing component can also be produced in a single operation. The backing section must be so highly compressed and solidified that it meets the requirements made for rigidity and strength. It must, simultaneously, be ensured that the tactile component retains its volume and, therefore, also its softness. In addition, a joint with the decorative material must also be created during direct pressure moulding of the backing and tactile elements.

Three types of polyester fibres are available as the basic materials: Bicomponent fibres (a) consist of a sheath and a core, the sheath material having a lower melting point and providing the joint between the fibres. The amorphous fibre (b) is a partially expanded amorphous polyester fibre which can be pressure-moulded only under simultaneous pressure and elevated temperature. The standard fibre (c) is a standard type used as an extender fibre.

Backing and tactile fleeces with various weights per unit of area are available as semi-finished products.

The backing fleeces are composed of the above-mentioned types of fibre, mainly at ratios of 40 % (a), 40 % (b) and 20 % (c) and have weights per unit of area of 200 to 2000 g/m^2, depending on requirements. Manufacture of the backing fleece is accomplished mechanically by means of needling, significant influencing variables being rate of material feed, penetration depth, penetration density and needle type.

Tactile fleeces consist mainly of 40 % (a) and 60 % (c) and have a weight per unit of area of 50 to 250 g/m^2. Manufacture of tactile fleeces is accomplished by means of thermofusion of the fibres. The connective-fibre-containing fleece is set by means of heat treatment without application of pressure in a continuous-type furnace.

The manufacture of a three-dimensional decorative mirror for a motor-vehicle side panel trim with the following structure is described below by way of example:

Fabric, leather or film, coated with hot-melt adhesive, are used as the decorative materials. A tactile fleece serves as the soft intermediate layer, while adequate rigidity is provided by a backing fleece.

All three components are placed in a pressure-moulding tool which is cooled on the decorative element side and heated on the opposite side to 205 °C (see Figure 9). Stabilization of the fleeces and of the decorative material is accomplished during thermal pressure-moulding. The temperature of the pressure-mould must be adequate for adhesion of the backing and tactile fleeces and must achieve reactivation of the hot-melt adhesive. The component can be removed after a cooling phase.

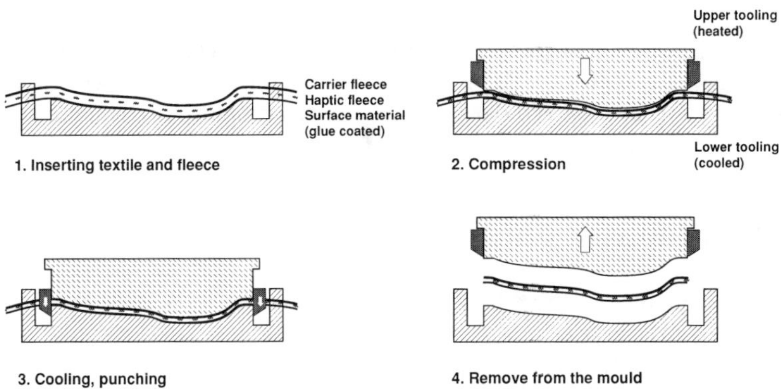

Figure 9 Pressure thermo-forming of fleeces (fleece moulding)

The cycle time for pressure moulding is approx. 30 sec, and the cooling period also approx. 30 sec.

This process, which has been developed to readiness for series use, is to be used in future vehicles. Further cost advantages can be achieved if the tactile element can be eliminated (as in a rear shelf, for instance) and polypropylene fleeces can, in addition, be used to provide the backing elements.

Blow-moulding technology

The blowing of hollow bodies on the basis of extrusion technology also offers, in the form of a low-pressure process, potentials in principle for the production of large decorated components. The basic problem in extrusion blow-moulding is that of harmonizing the continuous manufacturing process of the tubular with the intermittent blowing operation, which becomes the more complex when surfacing materials in the form of textiles or sheets are additionally to be positioned in the tool. Blowing technology is nowadays capable of producing even quite complex 3D geometries, but the process nonetheless still prefers one-dimensionally oriented (long) components where possible. In the case of interior trim components, therefore, this technology would appear to be less suitable for large rectangular door trims, for instance, but there are potentials for rational application in the case of rear shelves and similar long items.

Components notable for high rigidity (for a folding rear shelf in a three-door model, for instance) achieved thanks to their partial or complete double-skin design can thus be created. The low pressure and the lower melt temperature as compared to injection-moulding permit nondestructive back-cushioning of the surfacing materials.

One disadvantage is the insulating effect of the surfacing material on the cooling behaviour of the tubular (reduction of rate of cooling), with disadvantages for cycle times and therefore production costs. In addition, the application of the surfacing material on one side makes the manufacturing process significantly more difficult. Refinements in this field will indicate the extent to which this technology might provide serious competition for the other processes. It is certain that both the cost-effectiveness and recyclability of such a composite system require investigation.

Thermosets and back foarming technology

Of the thermosets, the polyurethanes which, as semi-rigid foam materials provide the link between the surfacing materials and backings produced, in each case in separate moulds, are worthy of consideration due to their comparatively low raw-materials prices. This long familiar technology is the most complex of the processes discussed. The extra complexity necessary in these foam components, however, permits the manufacture of items of an extremely high quality, practically unrestricted selection of decorative elements, adjustment of hardness (touch), and any backing. It is also possible to create for recycling reasons a single-type PUR composite consisting of decorative element, filling foam and backing. Economically viable recycling processes are then compression moulding, glycolysis processes and the manufacture of sandwich components.

Foamed components set the quality standard for the competing processes for the dashboards in quality vehicle categories.

In polyurethane technology, too, increasing price pressure is leading to the need to develop new rationally priced and competitive alternatives for interior trim components. PUR-RRIM systems make it possible to manufacture surfacing materials (sheets and textiles) nondestructively at low pressures and temperatures [5], Figure 10

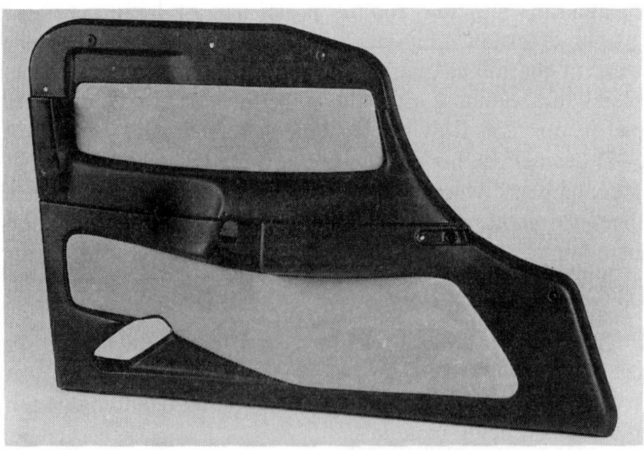

Figure 10 Door trim of the new Lancia Delta using the PUR-RRIM process

In this process, the preshaped decorative material is positioned in the lower half of the foaming tool. Inserts and other fixing elements can be placed in the upper half of the tool. After the locking operation, the short glass-fibre reinforced PUR-RRIM system can be incorporated at a low, decorative-element-preserving pressure level.

According to data supplied by the manufacturer [5], any surfacing materials can be used (given appropriate preparation). The tactile element must, in this process, be achieved by means of the foam backing of the sheets. A single-type three-layer material structure cannot be achieved.

Selection and influence of decorative materials

Despite the diversity of the individual low-pressure processes, the behaviour of the decorative material vis-à-vis the backing material to be processed in the mould is a problem in principle. The most varied combinations of melt pressure, melt temperature, wall shearing stress and mould temperature exert on the decorative material loads which it is required to resist.

Textiles

Figure 11 shows a typical structure for a decorative material. It consists of the textile decorative material in the visible area, the surface material, a compressive flexible intermediate layer and the backing material below this. The surface material generally consists of PES or PA textiles and is responsible for the visual impression presented by the component and for its "haptics" (subjective tactile sensation). The intermediate layer has up to now generally consisted of PVC or PUR foams, but could equally be produced in PES fleeces. It serves to achieve a "soft touch", and also as the thermal insulation between the hot-melt and the surface material (this layer can also be omitted when a "soft touch" is not absolutely necessary in a component). The backing material takes the form of a knitted fabric, knitted material or fleece which is, on the one hand, also intended to contribute to thermal insulation and, on the other hand, to protect the upper layers against penetration by the backing material

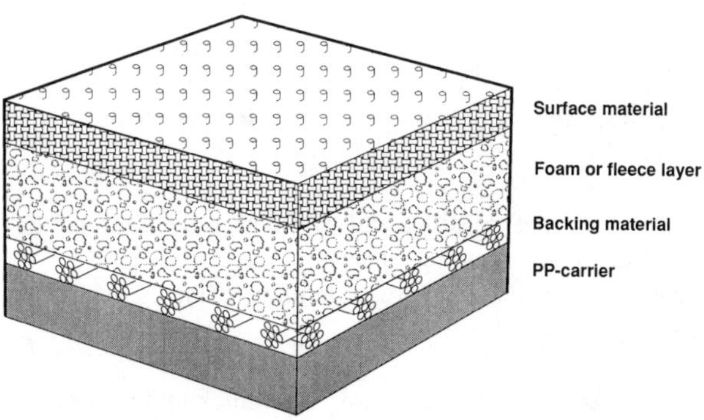

Figure 11 Schematic of the decorative element structure

and prevent the destruction this would cause. In addition, the backing material is required to provide a connection between the textile and the plastic backing by means, for instance, of fusion or by means of the penetration of hot-melt.

The high flexibility of the decorative element structure is of great significance for the avoidance of wrinkling, particularly at seams. Another important factor is the compressive flexible behaviour of the soft intermediate layer which must be retained during processing despite the internal mould pressure. Corresponding test apparatus for assessment of draping behaviour and of permanent deformation have been tested at the IKV, Aachen [6]. Our own studies have demonstrated that adequate adhesion between the decorative element and the backing material can be achieved without difficulty using a knitted fabric or fleece as the backing material.

Sheet and film

Both single-layer and double-layer ABS/PVC and PVC sheets with a foam backing consisting of PUR or PVC foam are generally used for laminating. In the case of the sheet for decoration of trim components using low-pressure processes, these are a double-layer structure consisting of a compact top sheet and below this a layer of foam. ASA and TPO sheets are the main alternatives for future applications in view of the requirement for avoidance of PVC and the need for the best possible recyclability. ASA sheet at present retains a certain development lead. Several sheet manufacturers are working on the refinement of the TPO sheets and clearance for series manufacture of these sheets can also be anticipated in the foreseeable future.

The sheets are manufactured by means of extrusion or calendering; the leather-like grain is achieved by means of an embossing roller.

The polyolefin backing foams used for decoration of low-pressure process moulded components were originally developed by two Japanese companies which, according to information from Toray, use differing processes for the production of PP foam (on the one hand, the complex salt-bath foaming process, achieving an extremely uniform cellular structure and, on the other hand, vertical foaming, with cells of varying sizes). The foam sheets produced in the salt-bath exhibit advantages for processing, particularly where high degrees of deformation are necessary. European products with comparable positive properties have now become available.

The PP foam is produced as rolled good at various densities and is joined to the compact top sheet by means of thermal laminating or flame bonding. Thermal laminating is an extra operation and includes a certain erosion of the foam, whereas flame bonding is conducted simultaneously to the embossing of the sheet.

The upper sheets are only 0.35 to 0.5 mm thick whereas the foam has a thickness of between 1 and 4 mm, depending on requirements. The finished foam sheet is thermo-formable; adhesion to the PP backing is achieved in the low-pressure processes by means of melting on of the foam with no adhesion promoters or adhesives whatsoever. Because of the materials' differing polarities, no adhesion with ABS is possible [4].

For moulded skins and compact foils, such as are used in in-mould skinning with PUR and, in some cases, laminating, attention must be devoted when using the low-pressure process with thermoplastic backings, as in the case of the two-component injection-moulding process, with which materials can form a joint (e.g. ASA foil with ABS).

Leather

Leather can, in principle, be processed in a similar manner to textiles with a tactile element and, possibly, a dividing layer. Since leather is significantly more resistant to penetration, however, it may in individual cases be possible to dispense with a multiple layer structure. In such a case, the leather will, in addition to sealing of seams by means of adhesive tape or similar, only require hardening against penetration with hot-melt. A special emulsion which seals leather and tactile foams applied to it against liquid plastic compounds and, in addition, permits single-type separation of decorative elements and backings for the purpose of recycling, is currently under test at Haas.

Table 2 Comparison of the characteristics of the low-pressure processes

Process / Characteristics	THERMOPLASTICS							PUR
	LP injection moulding	Gas injection moulding	Vertical injection moulding	EPP	Propellant type injection moulding	PES Fleece moulding	Blow moulding	RRIM
Compression load on composite	—	—	+/-	++	+/-	++	+/-	+
Thermal load on composite	—	—	+/-	+	—	+	—	++
Rigidity of composite	+	++	+	– –	+	- -	++	+
Lightweight potential	+/-	+/-	+/-	+	—	+	+/-	+/-
Fracture behavior of composite	+	+	+	++	—	++	+	+
Acoustic insulation properties	+	+	+	—	+	—	+	+
Linear thermal expansion	+/-	+/-	+/-	—	+/-	—	+/-	+
Cost-effectiveness	+	+	+	—	—	—	—	—
Overall score	+/-	+/-	+	+/-	+/-	+/-	+/-	+

++ very good
\+ good
+/- neutral
— poor
– – very poor

The recyclability of composite systems

The broad variety of materials is the greatest hindrance to the recycling of interior trim components.

Above all, the heterogeneous composite systems, which can be reduced back to their individual materials either not at all or only with great effort, scarcely offer any prospects for cost-effective, and therefore viable, recycling. Even thermal recycling encounters resistance when, as is generally the case, the composite systems contain PVC or PVC-containing materials. For this reason, a future objective must be the definition of materials combinations which are compatible, single-type wherever possible, and halogen-free, for future interior trim composite structures.

Which systems might be suitable for this purpose?

Surfacing materials	Haptic materials	Backing materials	Recyclability
PP sheet PP textiles	PP foams	PP modifications	Materials/ PP injection-moulding
ASA sheet	--- (PP foams)	ABS ABS/PC	Materials/ ABS injection-mouldings; problem: PP foams
PUR skins TPU skins	PUR foams	GF-PUR backings	Materials/ or chemical
PES textiles	PES fleeces	PES fleeces	Materials/ textile or injection-moulding

In principle, the polyolefins and polyurethanes offer the best prospects for single-type composite systems with sheet-like surfaces. The ABS systems suffer from the deficiency of there being no compatible foams.

Single-type polyester structures are suitable only for those components with textile surfaces. Since single-type PUR systems cannot be created cost-effectively in a single tool on a three-layer structure basis, only the polyolefins will be left as the potential solution as the cost pressure increases on the new generations of vehicles.

Recycling tests on interior trim components with textile polyester surfaces in some cases have produced positive results where the greater part of the textiles has been removed by means of wind sifting. A remainder of textiles can be left in the polypropylene.

The technological and economic properties of the low-pressure processes

Table 2 shows a summary of the characteristics of the low-pressure processes (using the example of a motor-vehicle interior trim component). There can be no doubt that the result of the analysis depends on an extremely large number of factors; in particular, the component to be manufactured plays the dominant role, as a function of geometry, boundary conditions and requirement profile.

For this reason, slight differences in the results may occur in specific cases on a subjective analysis.

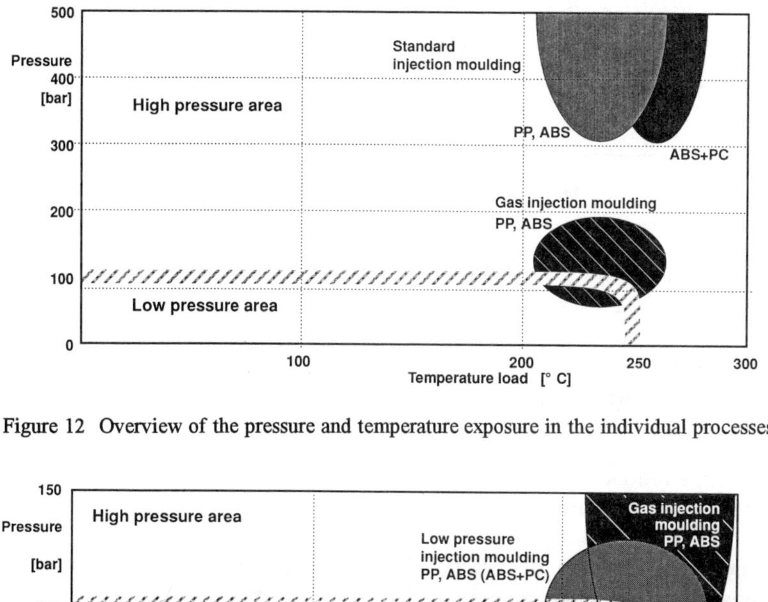

Figure 12 Overview of the pressure and temperature exposure in the individual processes

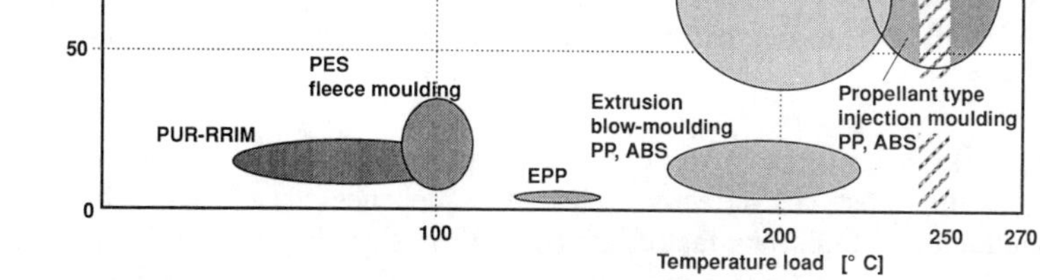

Figure 13 Low-pressure range pressure and temperature exposure

The basis for the analysis is provided by the pressures and temperatures to which the decorative materials are exposed, as shown in Figures 12 and 13. Table 3 provides an approximate overview of the physical properties of certain widely used materials.

Table 3 Physical properties of the materials used in low-pressure processes

	Modulus of elasticity in flexure	Impact strength (unnotched)	Elongation of break	Tensile strength	Density	Coefficient of linear thermal expansion
	N/mm^2	kJ/m^2	%	N/mm^2	g/cm^3	10 E-6/K
PP (with talkum and EPDM modification)	1400-2800	40-50	30-140	20-35	0,98-1,05	80-120
ABS	1800-3000	65-75	10-25	30-45	1.05	60-110
EPP	35-40	---	10	1	0,08-0,1	56 - 88
PET-Fleece	800 (Tensile)	---	13-18	16-19	0,6	0
PUR-RRIM	900-1250	10-16	4-10	13-20	0,6	22-35

In principle, however, two processes are particularly conspicuous as a result of their balanced characteristics profiles:

- vertical injection moulding using polypropylenes and
- back foaming using PUR-RRIM systems.

Both processes permit the rational production of high-quality interior trim components, the in-mould embossing method exhibiting the greater potential for cost-savings.

In Japan, door panel trims and simple dashboards have been manufactured for some considerable time now using the VIM process, the decorative sheet consisting of PVC, and the soft foam and backing of PP (see Figure 14). As already noted, the economic advantage can be found principally in the potential for manufacture of the moulding in a single operation. Complex and expensive operations such as the spraying of the separately manufactured backing with an adhesive and subsequent pressure or vacuum laminating are eliminated.

It is not possible to make a universally valid statement as to which of the low-pressure processes described is the most cost-effective, since this depends very heavily on the component to be manufactured. Important factors are component size, degree of deformation, selection of the decorative element (with a soft touch) and the design of the overfold. The costs for certain processes are compared, using the example of a rear shelf, in Figure 15. This is a large, relatively flat component with a textile surface which makes no particular demands in terms of softness, and with a overfold of 90°.

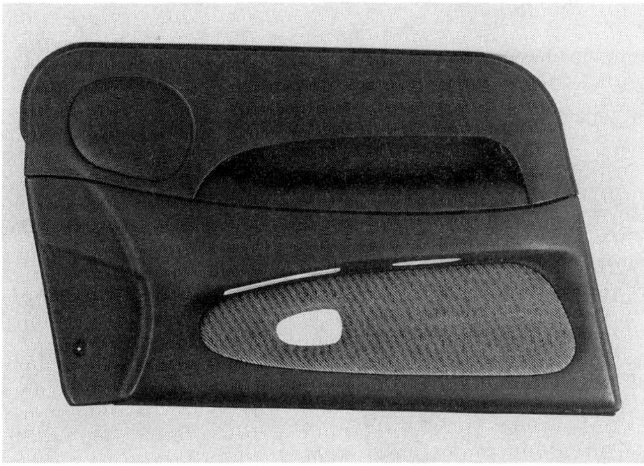

Figure 14 Door trim of the Mitsubishi Lancer, VIM process

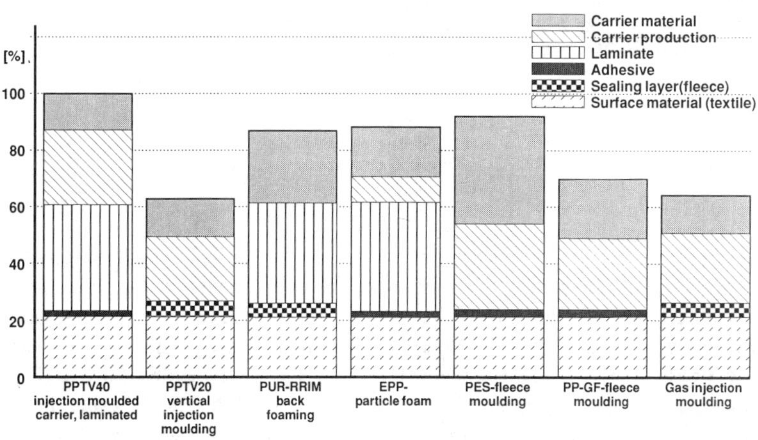

Figure 15 Relative costs for various low-pressure processes using the example of a rear shelf

For recycling reasons, intensive work is currently being devoted to replacement of PVC decorative sheets with halogen-free alternatives. The potential choices are ASA and TPO sheets. Materials continuity even in multi-layer mouldings will, however, be possible ultimately only on the basis of polypropylenes.

Tests in BMW AG's plastics and textiles division have demonstrated that the recyclate obtained from such comminuted VIM components can, up to a certain percentage, be added without difficulty to the manufacturing process as a substrate material. Punching and cutting waste from the production of new components can therefore be entirely integrated into the ongoing manufacturing process or used for other components.

The low-pressure processes also make it possible to create mixed surfaces in a single tool. Door trims with a sheet surface and a textile inlay can, for instance, be produced cost-effectively in a single processing operation; both sections can, in this case, also be equipped with a soft-touch (see Figure 16).

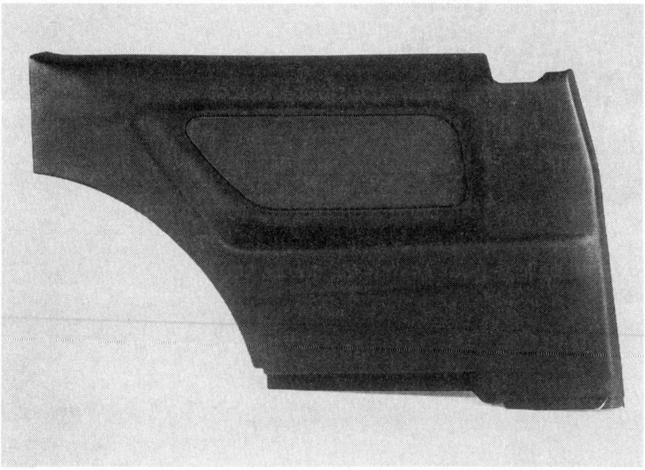

Figure 16 Rear quarter panel with textile mirror, VIM process

Limitations in the low-pressure processes are apparent in the overfold at the component edges. Overfold angles of 90 to 120° can be created on all sides of the component without difficulty. It should be noted for processing of foam sheets that only the underside of the foam backing undergoes the complete overfold and that the decorative element has a somewhat slighter overfold as a result of the shear strength resistance of the foam. Larger overfold of up to 180° have already been produced on one component edge, but 180° overfolds on all sides have been achieved up to now only on specimen components on a laboratory scale and not, as yet, in series production.

PUR-RRIM technology offers better prospects in this context, since edge thickness increases can be achieved with fewer problems than in the case of the thermoplastics. There is, nevertheless, a possibility of producing a 180° overfold using in-mould embossing technology, either in the same mould by means of an intelligent mould system, or in a downstream station.

The problem of undercutting occurs if this overfold is to be achieved in the mould. A flute positioned in the third plate is responsible for forming the overfold [7]. In order to permit creation of a moulding using an embossing process, the mould must be equipped with a shear edge. The decorative element would first be fixed in the separation level between the third plate and the die-side. The mould would then move to the embossing position, the third plate then being pressed against the nozzle side by means of correspondingly large springs. This would prevent leakage during injection and the embossing phase (see Figure 17).

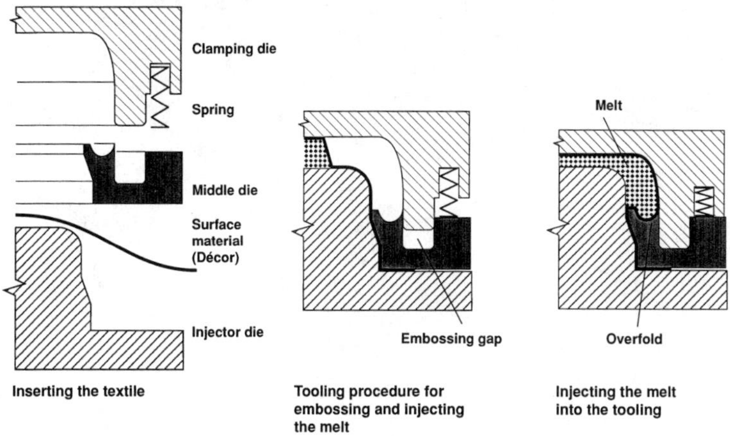

Figure 17 Mould concept for production of an integrated overfold, VIM technique [7]

If the integration of a bead system into the mould is not possible because, for instance, of an excessively complex geometry resulting, there is an equally adhesive-free alternative employing a downstream overfold system, such as is used after the laminating process. Adhesion is achieved in the PP structure not, however, by means of an activatable adhesive but instead by means of the linear melting-on of the thermoplastic backing by means of hot air or infrared heating.

If it should prove possible to implement this processing stage for large-series production, further cost-saving potentials and an unrestricted range of applications in the vehicle interior will result for the low-pressure processes (see Figure 18).

Prospects

For the future, productivity must be further improved for motor-vehicle production and supply-industry applications. The low-pressure processes provide an opportunity of creating an advantageous starting basis for the new generations of vehicles by means of product and process innovations. Customer-relevant characteristics such as finish quality, visuals and tactiles can certainly at least be retained if not actually improved. Such innovations can be most advantageously implemented in terms of quality, costs and time if the entire value-creation progression is performed by a highly capable and efficient supplier. The market no longer demands individual part-solutions but rather complete modules provided by system suppliers.

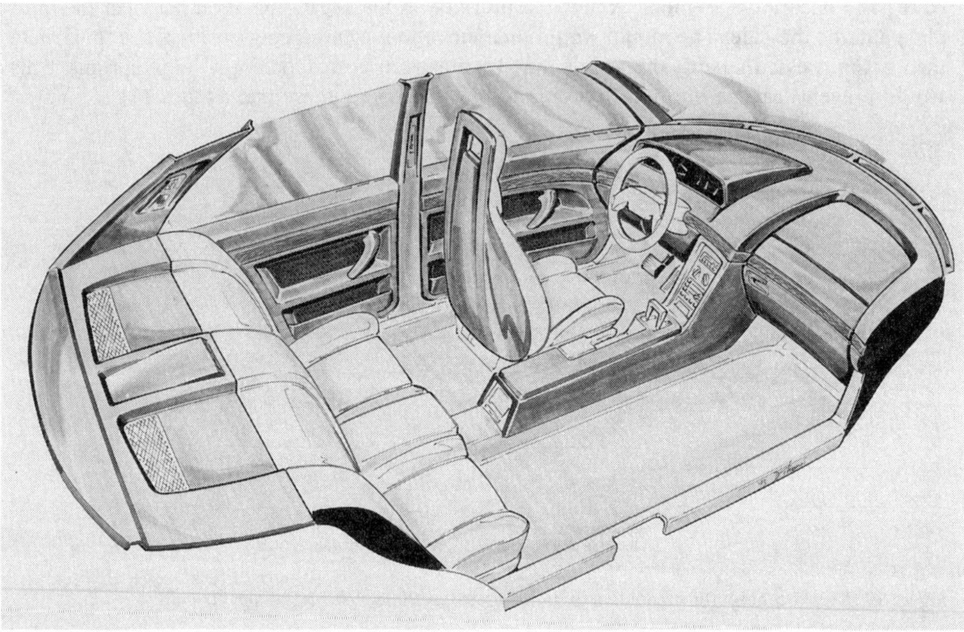

Figure 18 Process-oriented styled vehicle interior

The process-oriented and materials-oriented design of the components must, therefore, start not later than the styling-phase for the interior trim. A capable supplier will be able to perform conceptional and design work at the development stage as a function of internal synergies between the individual function sectors in such a way that the needs and exigencies of micro-economics, tool-making, production planning, production and, not least of all, quality assurance, can be taken into account.

The high demands made by the market will, in the future, too, challenge both the motor-vehicle industry and the supplier industry.

An equitable cooperation basis, characterized above all by

- trust
- continuity
- reliability
- clarity
- fairness and
- creativity

will make it possible to master these future challenges with high efficiency.

References

1 Eyerer, P., Bürkle, E.: Niederdruckspritzgießtechnik´(Low-pressure injection-moulding), Manuscript, 1991

2 N.N.: Zwei getrennte Fertigungsschritte in einem: Textil-melttechnologie (Two separate production stages in one: Textile-melt technology) Plastverarbeiter No. 44/1993, pp. 80/82

3 Eyerer, P., Bürkle, E.: Spritzgießen mit reduzierter Schließkraft (Injection-moulding using reduced closure force) Kunststoffe No. 81/1991, pp. 851/862

4 Bürkle, E.: Hinterspritzen und Hinterprägen von Textilien bei der Innenausstattung von Automobilen (In-mould decoration and in-mould embossing of textiles in the interior trim of automobiles) Düsseldorf: VDI Verlag 1992

5 Strapazzini, K.: A new competitive integrated solution for covered automotive interior trim parts, SPI Conference, Vancouver, 1993

6 Michaeli, W., Galuschka, S.: Hinterspritztechnik Teil 1: Eine Analyse der Randbedingungen (In-mould decoration, Part 1: An analysis of the boundary conditions), Plastverarbeiter 44/1993, No. 3, pp. 102/106

7 Michaeli, W., Galuschka, S.: Hinterspritztechnik, Teil 2: Eine Analyse der Randbedingungen (In-mould decoration, Part 1: An analysis of the boundary conditions) Plastverarbeiter 44/1993, No. 4, pp. 62 to 68

Steel chassis hybrid construction method: Aluminium/plastics body on the S 300 NC urban service bus

Herbert Rösch

Introduction

Kässbohrer took up the development of low-floor vehicles at a very early stage. The principal objective was always the achievement of the best possible vehicle concept from both the user's and the operator's point of view. The four main requirements for a municipal omnibus were always emphasized:

- maximum spaciousness for a given vehicle size
- best possible space utilization by means of largely flat floor surfaces
- fast passenger boarding and alighting
- cost-effective operation.

Various new technical developments affecting the suspension and drive train and the base frame have made it possible to achieve an extremely low-floor design. Various materials are given a load-bearing function for the first time in omnibuses.

Base frame	Steel
Superstructure	Aluminium
Front and rear assemblies	Reinforced plastics

The concept of this vehicle is elaborated and the most important individual technical solutions discussed in more detail below.

Figure 1 S 300 NC low-floor bus

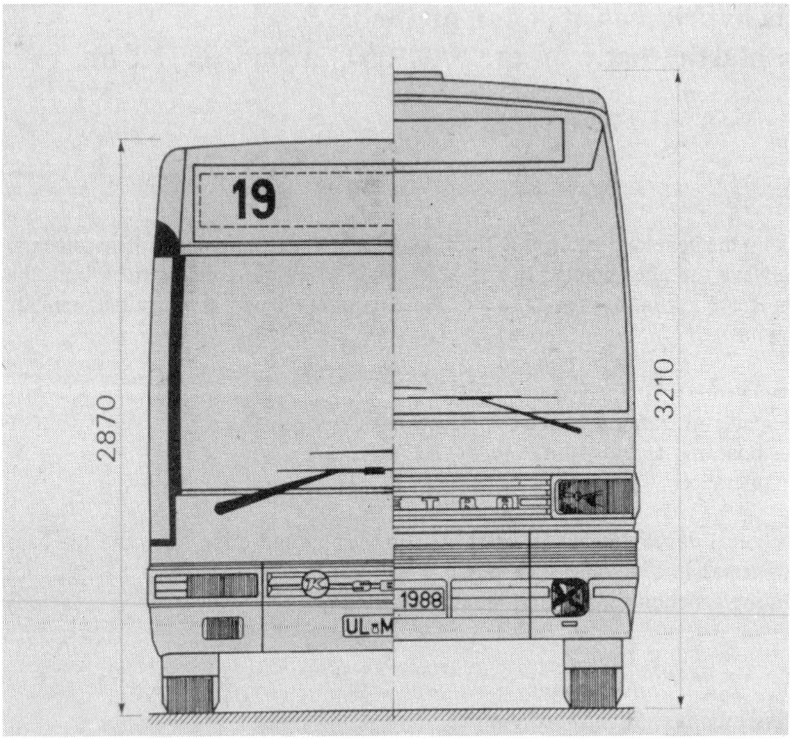

Figure 2 Front group/cross-section compared to conventional S 215 SL municipal omnibus

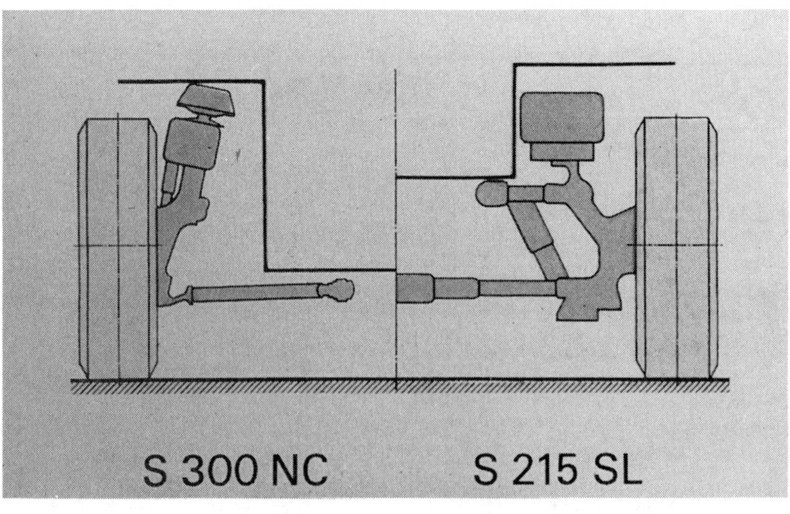

Figure 3 Foot space with strut suspension as compared to conventional design

Overall technical concept

Development work placed great emphasis on a visually highly attractive exterior and interior for the vehicle. The exterior can be seen in Figure 1.

Figures 2 and 3 illustrate the vehicle's discernible difference vis-à-vis its predecessor.

- Overall vehicle front-profile significantly smaller than the preceding S 215 SL model, see Figure 2
- the new axle design employing the McPherson strut principle makes the foot space shown here possible. Also clearly apparent is the improvement on the double wishbone suspension system used up to now, see Figure 3. Figure 4 illustrates the principal features of this vehicle type.
- Low floor in the front section of the vehicle, with no-step boarding at Door I and Door II
- Three double-width doors
- Standing area opposite Door II and Door III.

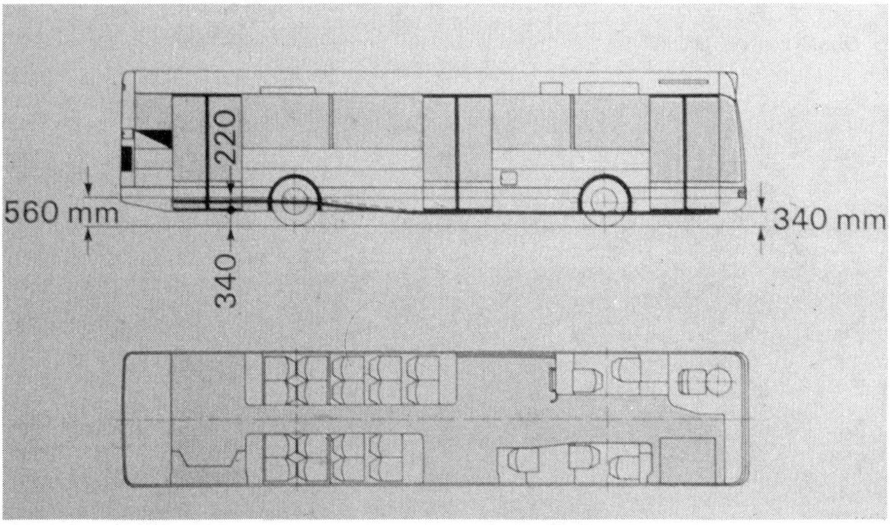

Figure 4 Overall concept: seats and aisle

Figure 5 shows the vehicle concept with ramp near Door I. The section through the vehicle shows the standard version, with capacity for 23 passengers seated and 81 standing.

Basic superstructure system

Like all previous Kässbohrer vehicles, this bus is designed as a tubular lattice structure, see Figure 6. The essential difference is the fact that new materials have been used.

- Base frame in steel
- Side panels and roof in aluminium
- Front and rear assemblies in plastics.

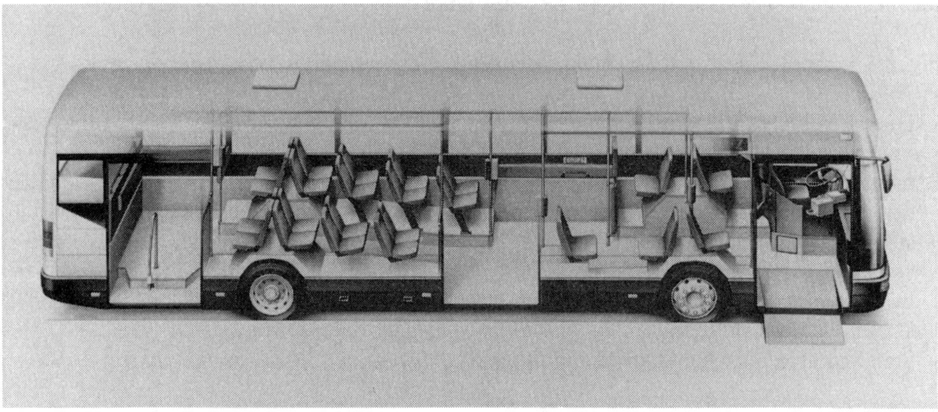

Figure 5 Overall concept: Interior view

Figure 6 Lattice framework structure with beam characteristics (FEM view)

Immediately following definition of the basic concept, the vehicle structure was mathematically reworked using the Finite Element Method (FEM). Load-oriented dimensioning of the structural elements, with appropriate allowance for the relevant load cases, was therefore possible as early as the conception and design phases. Strain measurements were performed on the completed vehicle's base frame and aluminium superstructure for final confirmation of the results of the simulation. Thanks to materials-oriented dimensioning (differing modulus of elasticity of steel, aluminium and plastics), the permissible stresses were never exceeded. In accordance with the SETRA (from the German term selbsttragend = self-supporting) construction principle of our company, the superstructure and base frame both make equal contributions to the overall strength of the vehicle's body.

Figure 7 shows a principle diagram of the base frame. The points indicated are the bolted joints to the aluminium superstructure.

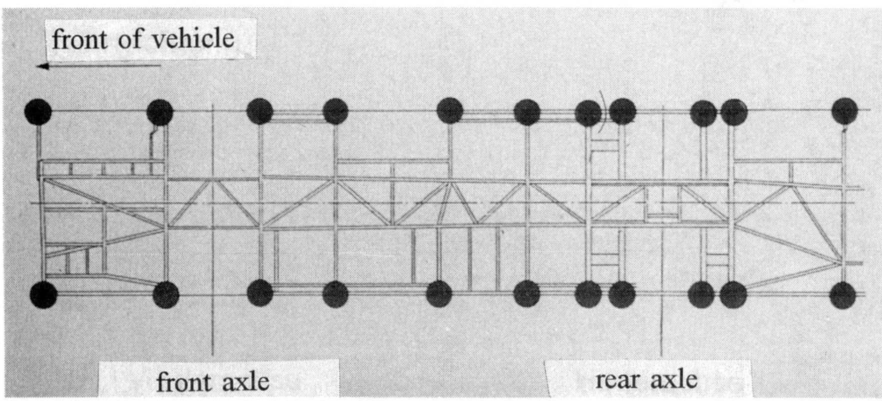

Figure 7 Connection of aluminium superstructure to base frame

Aluminium superstructure

The superstructure was elaborated on the basis of cooperation with Alusuisse (CH) from newly developed aluminium extrusions using the M 5438 system. The individual extrusions can be assembled extremely rationally using a special bolted joint. The design principle can be seen in Figure 8. The side panels and the roof are first mounted in a simple jig and then joined to one another. Purely bolt-based assembly of the ready-shaped sections achieves shorter manufacturing times. The superstructure can, therefore, be assembled at diverse locations, including export-customer countries, for instance, simply by supplying the corresponding sets of parts. All the aluminium extrusions feature a tapered sectional groove. The sections are therefore precisely aligned with one another when joined using the connecting elements, i.e., the system is self-checking and self-correcting. During the bolt-assembly operation, the bolts are secured against slackening by means of a micro-encapsulated adhesive. Figure 9 shows a complete pre-assembled roof unit, with its full internal equipment, such as insulation and electrical wiring. The pre-assembly of certain major items, such as the roof assembly, for instance, makes it possible to achieve an optimal assembly cycle. Overhead and other ergonomically awkward work can also be avoided in this way. After a comparative appraisal of welding and bolting for joints, Kässbohrer decided in favour of the bolting system. This was the origin of the ultimate bolted-type corner joint, see Figure 10. For our particular application, the bolted corner joints achieves significant advantages in terms of dynamic load-bearing capacity (see Figure 11).

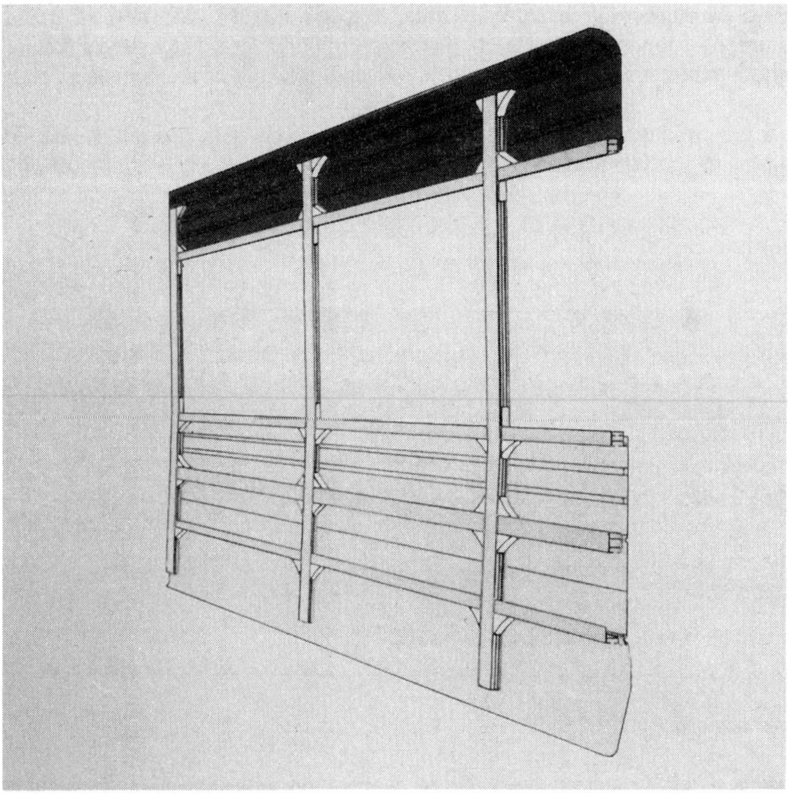

Figure 8 Aluminium superstructure

It was possible to achieve weight-savings of approx. 25 % as compared to the weight of a conventional steel superstructure. This equates to a weight reduction of around 250 kg, with no sacrifice in strength. Customers' requirements for greater comfort and customized interiors can thus be met without exceeding legally stipulated maximum axle loadings.

Front and rear modules as self-supporting plastic constructions

In both formal and functional terms, a new front group was developed in order to meet the requirements made on a modern low-floor bus. The front panel module takes the form of a self-supporting all-plastics structure of highly integrated design. The result was a weight-optimized easy-to-assemble and corrosion-resistant unit. This element is consistent with the overall production philosophy of very extensive prefabricated assemblies and structural components. The decisive factor in the selection of this concept were the relatively low tooling and component costs. Figure 12 shows the complete self-supporting design front end of the bus. The component is tested in its installation attitude on the 3D tester.

Figure 9 Preassembled roof unit

Manufacturing processes for the prototype and pre-series buses

The prototype and pre-series production run are created from pigmented GF-UP 30 using the hand assembly method. The exterior has a fine sandable coating, while the interior has a fibre texture. Material thickness is 3 to 4 mm. The front of the bus consists of a total of eleven individual components. Of these, nine components are produced in the form of single-shell GRP laminates. The components are trimmed, the matching adhesive surfaces are fitted up, and the surfaces roughened. After alignment of the individual components, all eleven individual items are fixed in a special bonding jig and adhesively bonded to one another (see Figure 13). Bonding is followed by a further CNC trimming operation for the entire front assembly of the bus. All the holes necessary to house the female-threaded rivets used are also made during this operation. The stud mounting for subsequent bolting on of the route destination indicator box is fitted to the framework roof structure prior to adhesive bonding. The AO pillar is produced using an injection process, a pre-equipped PU foam core with a weight per unit of volume of RG80 being enclosed in several layers of tubular fabric. A special glass/carbon-fibre tubular is used to absorb the enormous compressive forces in these zones (see Figure 14). The adhesively bonded joints are then optimized and the surface of the front assembly of the bus prepared for subsequent on-line painting in the assembly shop. For the series production run, the individual components are made of pigmented GF-UP material using the wet moulding process (see Figure 15). The interior components are as smooth as demoulded. Kevlar inlays are used in the case of critically loaded items, such as the window aperture, for instance. Local reinforcement in the form of glass-fibre fabric additions are provided around the fitted female-threaded rivets for the bolted joints (see Figure 16).

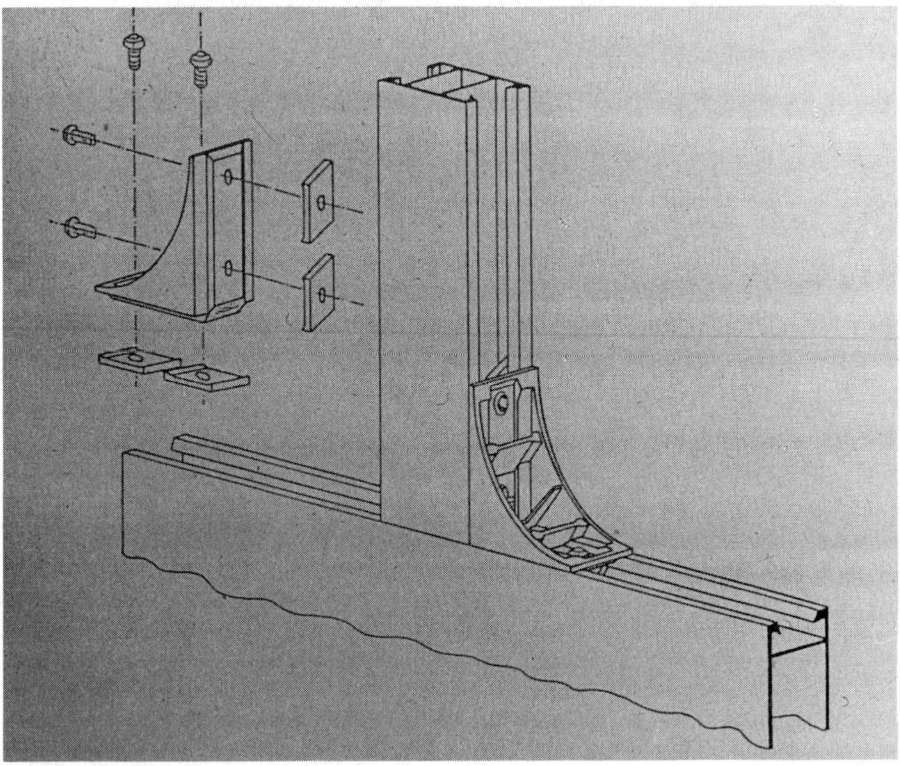

Figure 10 Bolted corner joint

Materials

Resin system

Palatal A410 is used for this application. This is a medium-reactive isophthalic acid polyester resin dissolved in styrene. This material is notable for its good thermal resistance and extremely high mechanical strength.

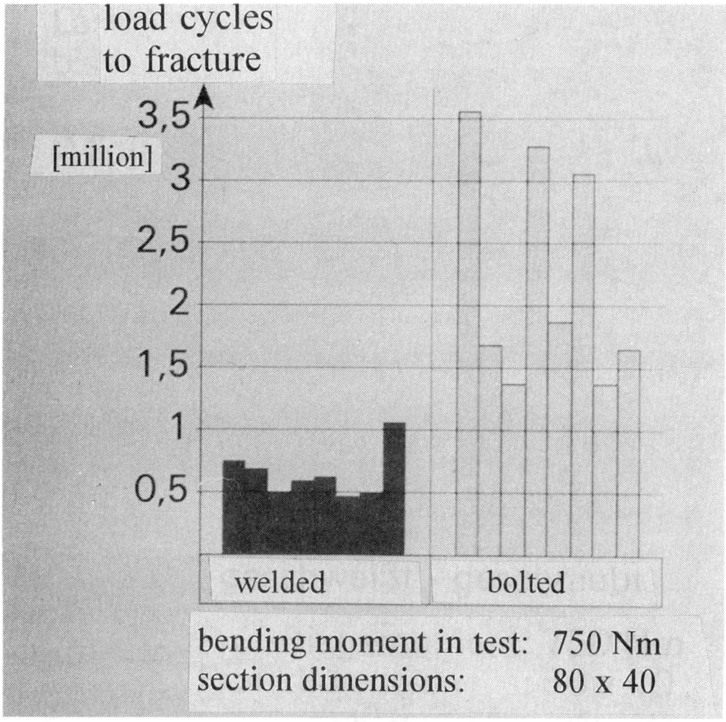

Figure 11 Corner joint strength appraisal: welding v. bolting

Glass

Basic structure consisting of textile glass mat of 300 or 450 g/m^2; geometrical shapes in cut non-oriented low alkali spun borosilicate glass (E-glass) fibres with plastic-fibre sizing. All load-bearing components are reinforced with 580 g woven glass-filament fabric. Surrounding the front screen, a layer of 25 to 30 g/m^2 Kevlar fabric, carbon-fibre/glass-fabric tubular for the AO pillar.

Adhesives

A two-component PUR adhesive is used. The component-dependent pot time is determined and adjusted separately. The adhesive is applied using a two-component system.

Bolted joints

Fully-automatically inserted female-threaded rivets are integrated for subsequent fixing of the component to the vehicle. These female-threaded rivets have M6/M8 internal threads and a knurled head. The required turn-out torques are achieved in this manner.

Figure 12 Plastic front module (C.F. Maier works photo)

Testing

The component conforms to the required fire-safety tests and meets the splinter test. Testing is performed to the UTAC, MVSS302 and corresponding DIN standards.

Attachment of the component to the vehicle

The component (front end or rear end) is attached directly to the vehicle using the pre-installed female-threaded rivets. In Figure 17, the front end can be seen directly attached to the vehicle, i.e., to the base frame. The plastics front end is a load-bearing element in the basic superstructure. In this example, it absorbs forces from the roof and the side panels.

Figure 18 shows a complete preassembled vehicle with aluminium side panelling and plastics rear assembly. The rear assembly is additionally reinforced with a load-bearing steel-framework structure, due to the high forces which occur in this area.

Figure 13 Adhesive bonding jig for front module (C.F. Maier works photo)

Figure 14 Reinforcement using carbon fibre tubular (C.F. Maier works photo)

Figure 15 Mould for front module (C.F. Maier works photo)

Figure 16 Female-threaded rivets for bolted joint, top (C.F. Maier works photo)

Figure 17 Attachment of front module to base frame (steel)

Summary and prospects

The concept presented here represents Kässbohrer's first-ever use of new materials for load-bearing assemblies. The specific advantages of steel, aluminium and plastics can thus be fully exploited. The municipal omnibus is the ideal technology demonstrator, due to the low production runs. The adoption of this design principle for other vehicle models is at present under examination. Cost-effectiveness would, however, need to be significantly improved.

Figure 18 Complete assembly of aluminium body side with plastic rear end.

Reference

1 Dr. Hagen, H.; Stangl, G.; Schmadl, D.: Die Idee des Setra Concept Bus
Der Nahverkehr 1/1988

Further development of the Neoplan MIC into a series vehicle; 5 years experience of the first coach in fibre-composite plastics

Bob Lee and Rainer Bickel

Development background

The development of the omnibus concept of the future started at NEOPLAN with the unveiling of the first prototype in May, 1988, with the objective of reducing both the financial and environmental costs of the construction and operation of municipal omnibuses, see Figure 1.

Figure 1 MIC Metroliner in carbon design

A zero series consisting of fourteen vehicles, which were able to prove their capabilities in everyday traffic while on loan to customers, demonstrated that a cellular production method based on the joining of two half-shells produced using the hand lamination method is unsuitable and too expensive for series production.

Series production

The second step taken was therefore the subdivision of the cell into a total of nine individual parts, the separate prefabrication of which could also be accomplished at dispersed locations, see Figure 2.

These individual components are brought together and assembled in a bonding jig which we designed ourselves, see Figure 3. This operation, which takes about four hours, is manageable and permits production of one cell per day, providing the correct number of prepared individual parts is available. Tempering of the cell was in each case performed overnight in a tempering kiln specially provided for this purpose.

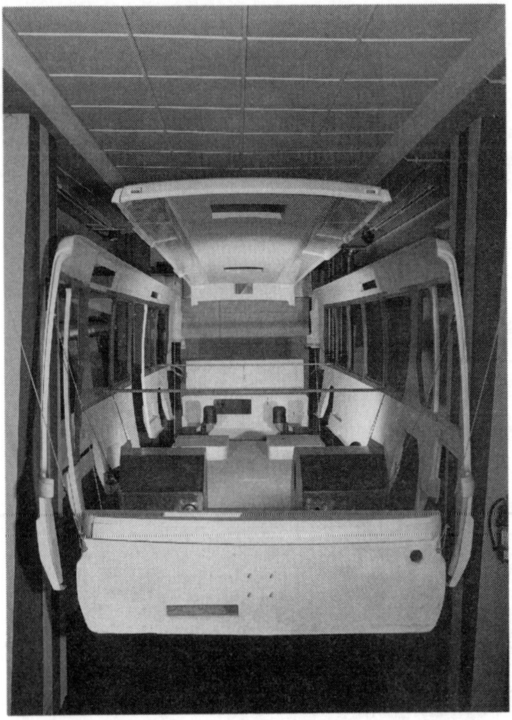

Figure 2 Shell parts

A further design provision on the route to series production was the optimization of the carbon-fibre content in the cellular structure.

In the course of revenue-earning service, it became apparent that the cells, produced as passenger vehicles in two sizes, 8.00 m and 10.60 m with an approved passenger capacity of 80 persons, are capable of fulfilling all the demands made on them.

The original requirements were:

- lightweight construction
- passenger friendly
- safe
- cost-effective and
- environmentally safe.

Four years of service, involving more than 160 such vehicles, have now proven the practical suitability of fibre-composite materials in this application.

Figure 3 Bonding jig

After problems at the interfaces between the steel components and the plastics bodyshell had been solved, difficulties then occurred as a result, mainly, of the cell's vibration behaviour. The plastics bodyshell reacts extremely sensitively to excitation by the diesel engine, suspension, etc. The strength advantage of the harmonious cellular structure also involves the creation of an ideal noise transmission element.

Here, too, we pursued new paths in the mounting of axles and rotary-vibratory components; the result of all these solutions is now perfectly capable of comparison with a steel-based vehicle.

The present-day development status was achieved by means of systematic reinforcement of heavily vibrating components, the use of technically optimized suspension elements and many other components tested in the course of exhaustive detail work.

The fibre-composite bus and the environment

The plastics bus had scarcely been unveiled when the environments began to question the recyclability of such a vehicle. Due to a lack of information, the broad mass of the public had reacted to the plastics vehicle with extreme reserve, since anything bearing the name "plastics" is immediately associated by the layman with environmental problems. Such prejudices, however, ignore the extremely positive effects which such a bus can exert on the environment throughout its life.

According to the literature available, the production of fibre-composite materials requires approx. 46 MJ of energy per liter of material produced. As a comparison to this, the energy involved in the production of a liter of steel is 450 MJ and for a liter of aluminium, a whole

640 MJ [1]. The energy savings achieved become even more impressive when one at the same time remembers that around three times as much steel (in terms of weight) is required for such a vehicle (see Figure 4).

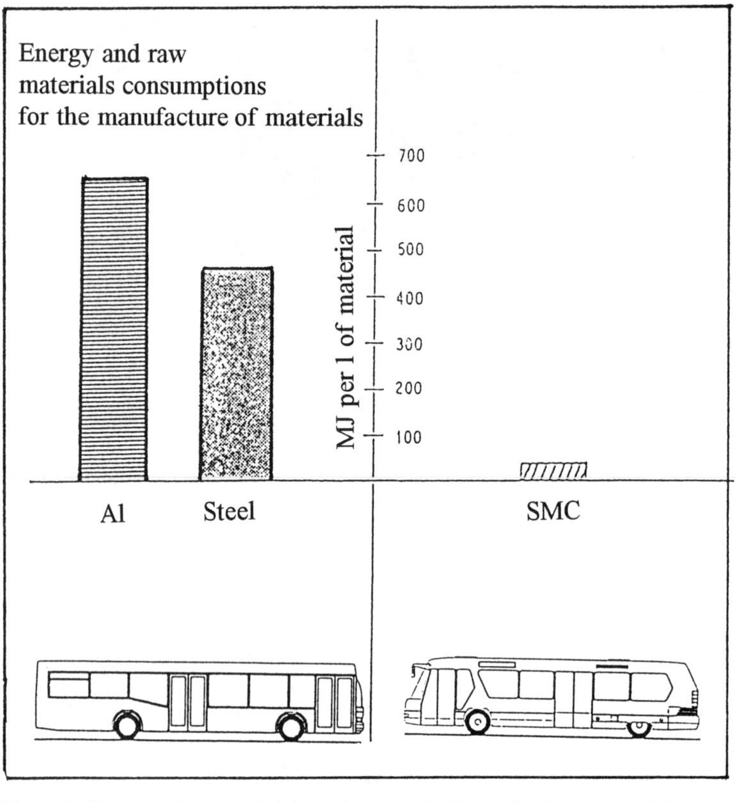

Figure 4 Energy and raw material requirements for the production of materials

The same also applies in the case of transport. Around 1 t of material must be transported in the case of the fibre-composite bus, while this figure increases to approx. 3.5 t in the case of a steel bus or a steel bus framework complete with steel panelling.

If one examines the way in which steel is produced, and the amount of oil or electrical energy required to smelt iron ore, create ingots of steel, convert these in another shaping process to flat material and convert this, again, in a further heating process into tubes, transport these by road from the steelworks to us and then cut them again by electrical means into smaller lengths, in order to immediately join them together into a steel framework, it then becomes clear how simple and straightforward the production of a bodyshell from fibre-composite materials really is. And even this account fails to make allowance for the fact that some 15 % of the steel material delivered is returned to the steel works in the form of waste and offcuts and is there again melted at a high energy cost.

A second and certainly no less significant saving is achieved during the operation of a plastics vehicle. Compared to a steel vehicle, some 3 to 4 t less mass requires moving, demonstrably reducing fuel consumptions by 25 to 33 %. If one assumes an average "mileage" of approximately 50 000 km each year, the result is a saving of around 7 500 liter, which, extrapolated for a twenty year bus service life, produces an extremely respectable saving of 150 000 liters of diesel oil. The energy contained in this fuel could be used to build a large number of new buses, see Figure 5.

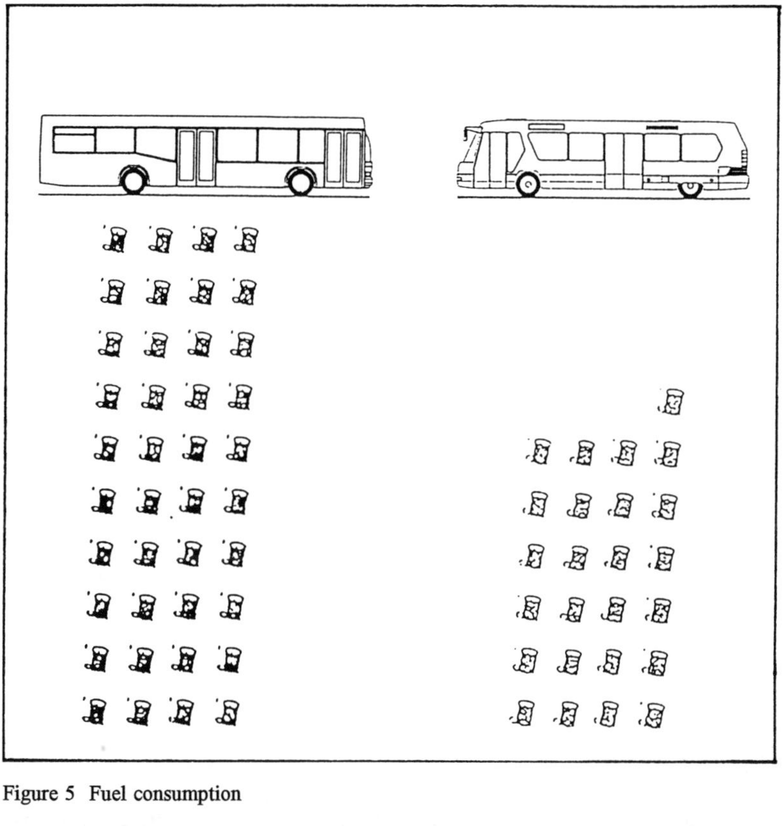

Figure 5 Fuel consumption

The lessening of the burden on the environment is also enormous, since these 150 000 liters of diesel fuel now no longer need to even be refined, and, because they are not combusted, there is less soot, CO_2 and NO_x produced.

The third environmental advantage derives from the fact that no special corrosion-prevention precautions need to be taken during manufacture. In contrast to a steel structure which, after assembly of the framework must be degreased either in immersion baths or by hand and then painted with protective coatings, this entire operation is eliminated when plastics are used. Cavity filling and sealing, which is necessary in all steel structures, can be entirely omitted. So there are no paints, no chemical reactions, and no resultant environmental pollution.

A further advantage can be found in the fibre-composite bodyshell's longer service-life, since it is immune to corrosion. A standard steel-bodied bus has a useful life of 10 to 12 years on average, whereas it can be assumed that a plastics bus will last for 20 to 25 years, see Figure 6.

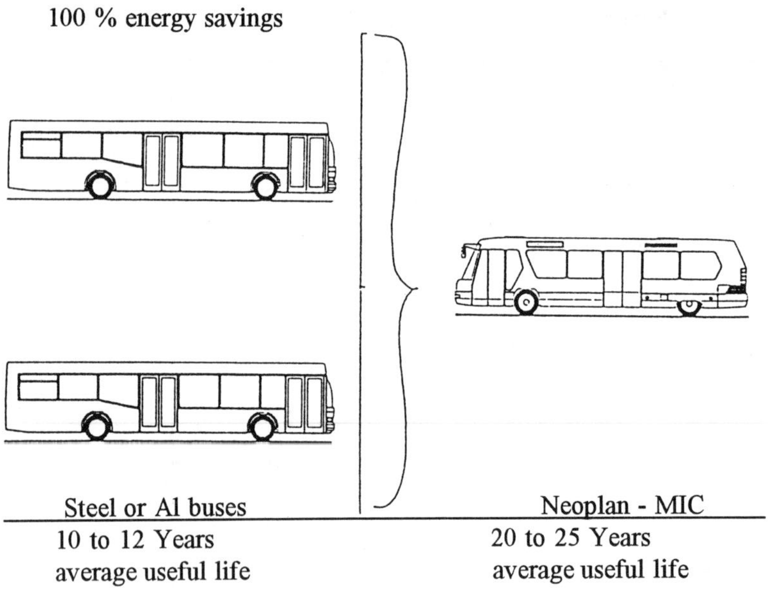

Figure 6 Period of use

So what happens to the plastics omnibus at the end of twenty-five years? We actually asked ourselves this question during the design phase. The axle suspension demonstrates how this patented system (we call it MFE = multi-function element) provides an ideally separable joint between the plastics cell and the attached steel components. This joint absorbs horizontal and vertical motion. The use of Celasto material enabled us to prove, in the course of a large number of tests, that, using this element, all the functions of the motions of the springs resulting from damping of the suspension can be absorbed without any wear occurring, see Figure 7.

The MFE elements make it possible to separate the steel components into single-type groups after their useful life without any detriment to the plastics structure. The service-life of these elements has been tested on hydropulse test stands, with a minimum of 500 000 to one million kilometers being indicated, see Figure 8. The complete drive train and rear axle takes the form of an assembly mounted on a sub-frame and connected to the bodyshell by means of six of the elements described.

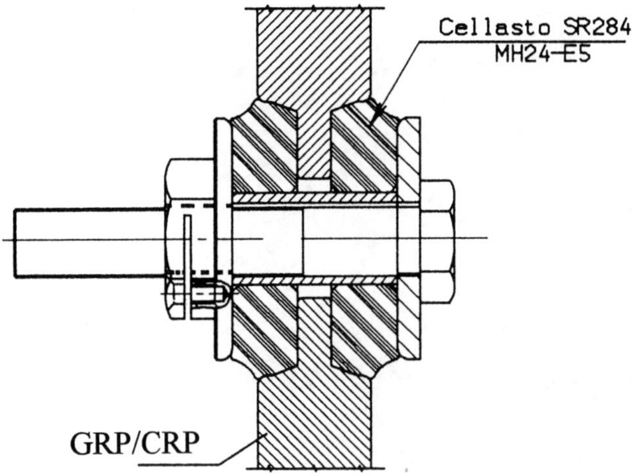

Figure 7
Multi-Funktions-Element (MFE)

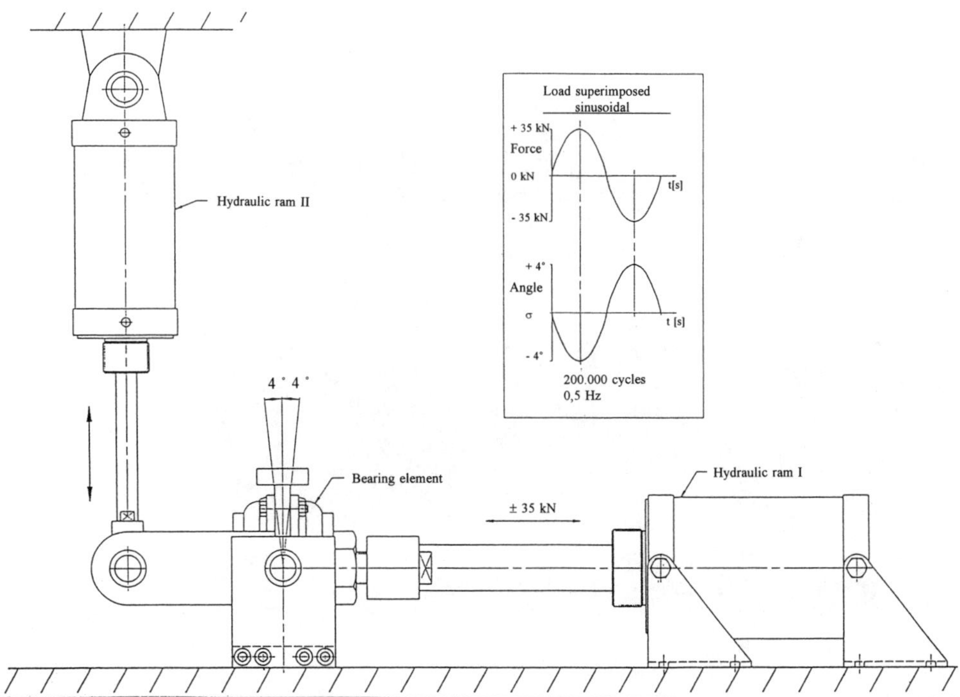

Figure 8 Test stand of MPE

This modular structure permits rational installation and changing of the engine. A diesel or an electric drive system can, for instance, be installed in the same plastics bodyshell, see Figures 9 and 10.

Figure 9 Electro cradle

Figure 10 Diesel cradle

At the end of our chain, therefore, we again have a single-material-type fibre-composite bodyshell. Such a material can be processed without difficulty by a waste incineration plant, for instance. It can, in fact, replace expensive oil which is added in order to achieve high combustion temperatures during the waste incineration process [2].

If we again examine the steel bus, which consists of highly diverse materials and also a very large number of fibre-composite material components riveted or bonded on, separation is significantly more difficult, costly and complex and, ultimately, recovery of the steel again incurs extremely high energy penalties. This must be set against the energy recovered from the fibre-composite cell in thermal disposal, which amounts to a further gain of 10 MJ per liter of material [1].

Time will tell which of the other technically feasible recycling processes is rational and cost-effective, and to what extents.

Market acceptance

The marketing policy which we operate initially envisaged supply of vehicles of this type only to Germany. The people concerned with research into new drive systems and smaller and lighter engines were naturally very interested in this fibre-composite omnibus. It is no coincidence, therefore, that the first diesel-electric power transmission systems are currently being tested in this vehicle.

Thanks to the vehicle's relatively low all-up weight, around 100 kW is sufficient for forward propulsion of such an omnibus under normal traffic conditions. This, in turn, means that all components can be considerably smaller than those in a standard municipal omnibus. This was also the reason for us exploring with the Magnetmotor GmbH of Starnberg the route of diesel-electric power transmission and the reason why this forms the basis for all new and future hybrid systems.

Irrespective of the source of electrical power, whether it be a simple diesel engine, a gas engine, a gas turbine or - in the near or even more remote future - even a fuel cell, it will always have been the lightweight omnibus which made such technologies viable.

We should note that worldwide interest in this vehicle has increased and that the fibre-composite bus is to be re-invented in the USA, too, by Northrop (the builder of the stealth bomber).

Operating experience

The power outputs of the new commercially available Euro I engines made it possible for us to replace the previous 230 h.p. six-cylinder engine by way of experiment with, inter alia, a 156 h.p. four-cylinder unit. Fuel consumptions dropped even further. We were able to achieve figures of between 15 and 17 liters for scheduled route operations in flat areas. This is slightly above one third of the consumption registered when using conventional technology. We benefitted, of course, from the new generation of engines and from the specially tailored high-performance automatic gearbox installed, but it must nevertheless be emphasized that none of this would have been possible in the form of an integrated system if conventional structural materials had been used.

A further application is our small electrical MIC. In this vehicle, the weight-savings compared to a steel vehicle of the same size are around 1500 kg, precisely the weight required if you want to equip a standard bus with batteries capable of permitting any kind of respectable range.

Here, too, it is apparent that the weight-savings initially achieved are brought up to the level of an in-service diesel-powered vehicle only when the batteries are installed. This has opened up a completely new path of development. The drive systems which we selected have been sufficiently specified to permit us now to be able to supply complete systems which, with a special battery-changing technique, permit 24-hour operation even of battery-operated electric buses.

Observation on economy

Our company last year transferred cell production to Prague, where series manufacture of the components for our plastics vehicles will continue in 1994. The lower wage costs in the Czech Republic permit us to transfer this manufacturing technology, which is still in its infancy. We are convinced that enormous steps forward will be made in this sector in the next few years.

Fibre-composite material bodyshells offer enormous potentials for the reduction of manufacturing costs, thanks to the potentials for integration of functions. The need for insulation, interior trim, etc., can thus be eliminated.

Bus-building is similar to boat-building on a somewhat larger scale, but can in no way be compared with motor-car or truck manufacturing. In our opinion, this is precisely the reason that fibre-composites are so eminently suitable for the construction of buses.

Our company intends to consistently pursue the route already embarked on and manufacture the highly promising combinations of the new drive systems paired with ultra-light plastics vehicle bodies.

References

1 BASF Publication: Neue Wege für Verwertung und Recycling (New Reutilization and Recycling Routes)

2 Menges, M. and Fischer, R.: Kohlenstoff-Recycling beim Aufarbeiten gemischter Kunststoffabfälle (Carbon recycling in the reprocessing of mixed plastics waste), "Kunststoffe" 81, Aachen 1991

Status report: Life-expired vehicle recycling and legal regulations - Current VDA/PRAVDA report
Status of life-expired vehicle recycling at Adam Opel AG

Peter Zumbroich and Bernd Kiefer

Introduction

In recent years, we have all become more aware than ever before of the consequences of irresponsible treatment of our environment.

The individual mobility provided by the automobile and the high standard of living which our economic system permits are inseparably linked to one another. This finds expression not least of all in the continually rising number of motor-cars in the European Union (EU), which has already reached a level of more than 142 million vehicles.

Topics such as the emission of pollutants, the choking of our cities with road traffic, noise, expenditure of precious resources and energy, and the environmental burden of disposal of life-expired vehicles have, on the other hand, become the subject of heated public debate.

These circumstances give rise to the need for the motor-vehicle industry to confront these problems and minimize the concomitant disadvantages.

One of the many associated difficulties is, without doubt, the no longer acceptable but still practised present-day methods of disposing of motor-vehicles once their useful life has finished.

More than 14 million life-expired vehicles were scrapped in western Europe in 1991; the figure for the Federal Republic of Germany (without the former GDR) was around 2.3 million. These life-expired vehicles are processed and disposed of for profit by a network of medium-scale scrap metal dealers. The problem is, however, that by far the major portion of these scrap dealers fail to fulfil modern environmental protection requirements.

As much as 75 % by weight of life-expired vehicles, composed practically exclusively of metals, are returned to the economic cycle via the current recycling route, namely the scrap-vehicle dealers and the next downstream recycling stage, the shredder. No other complex mass-produced product achieves such a high level of recycling. The remaining vehicle constituents - amounting to around 500 000 t each year in Germany alone - are still currently dumped at landfills in the form of so-called shredder lights.

It is worth noting in this context that the plastics content of shredder lights will also rise sharply in the future. It is predicted that the by-weight plastics content of shredder lights will increase to as much 40 % in the mid-term.

We may summarize by saying that, on a present-day view, recycling of life-expired vehicles is endangered by the following circumstances:

- generally shrinking landfill capacities,
- increasing costs as a result of the scarcity of landfill capacity and the greater demands made on landfill systems,
- increasing amounts of shredder residue material requiring dumping, due to increases in the use of plastics in each individual vehicle and a simultaneous rise in the numbers of vehicles being scrapped.

VDA - PRAVDA

In order to find a solution to the rapidly approaching conflict, the German motor-vehicle industry set up, as early as February, 1991, a work group bearing the name PRAVDA (Projektgruppe zur Altfahrzeugverwertung der deutschen Automobilindustrie = German Motor-vehicle Industry Project Group for Life-expired Vehicle Recycling) with the subject of life-expired vehicle recycling as its mandate. BMW, Ford, Mercedes-Benz, Opel, Porsche and Volkswagen/Audi are the members of PRAVDA.

PRAVDA deals both with the political background and the technical and economic aspects of life-expired vehicle recycling. The successful implementation of the concepts elaborated presupposes a broad level of consensus among all the branches of industry involved in recycling.

PRAVDA's objectives vis-à-vis the political world and the public are:

- the arousal of a positive public awareness regarding the motor-vehicle industry's capabilities and its will to achieve rational recycling, accompanied by awareness of the recycling successes achieved,
- education of politicians regarding suitable and rational objectives, strategies and provisions. The implementation and enforcement of rational concepts for motor-vehicle recycling.

The other operative objectives are:

- The setting-up (in cooperation with the other European manufacturers) of a recycling system which is intended to give the recyclers the potential (and the incentive) of recycling various German (and European) marques.
- Ongoing improvement (optimization) of motor-vehicle recycling with continuous reference to economic, ecological and other relevant targets.

The common motor-vehicle recycling concept:

The result of PRAVDA's activities up to now has been the modification of the VDA's "Joint Concept for Motor-vehicle Recycling" submitted as early as 1990 by the motor-vehicle industry, which was drafted on the basis of cooperation with the industrial associations affected and was based on market-economy principles.

The following confederations and associations jointly support this concept:

Verband der Automobilindustrie e.V.
(VDA = Confederation of the Motor-vehicle Industry)
Verband der Importeure von Kraftfahrzeugen e.V.
(VDIK = Confederation of Motor-vehicle Importers)
Zentralverband Deutsches Kraftfahrzeuggewerbe
(ZDK = Central Association of the German Motor-vehicle Trade)
Bundesverband der Deutschen Stahl-Recycling-Wirtschaft e.V.
(BDS = National Association of the German Steel-recycling Industry)
Deutscher Schrottrecycling Entsorgungsverband e.V.
(DSV = German Scrap-recycling and Disposal Association)
Verein Deutscher Metallhändler e.V.
(VDM = Association of German Metal Dealers)
Wirtschaftsvereinigung Stahl
(Confederation of the Steel Industry)
Wirtschaftsvereinigung Metalle e.V.
(Confederation of the Metals Industry)
Gesamtverband kunststoffverarbeitende Industrie e.V.
(GKV = Confederation of Plastics-using Industry)
Verband Kunststofferzeugende Industrie e.V.
(VKE = Confederation of Plastics-producing Industry)
Fachvereinigung Flachglasindustrie e.V.
(Technical Association of the Flat-glass Product Industry)
Gesamtverband Autoteile-Handel
(GVA = Confederation of the Automobile Component Trade)
Gesamtverband Deutsche Textil e.V.
(Confederation of the German Textiles Industry)

The market-oriented joint concept for motor-vehicle recycling is briefly described below using a series of headings, see Figure 1:

- Acceptance points for life-expired motor-vehicles consist both of motor-vehicle trade establishments and recycling establishments certificated by the motor-vehicle industry.

- These points issue the last owner of the vehicle with the so-called recycling certificate, submission of which to the registration authority is the precondition for the discontinuation of payment of tax for the vehicle.

- The last owner of the vehicle and the recycling establishment freely negotiate the residual value of the life-expired vehicle on market-economy criteria.

- A body in which all parties involved, i.e., the motor-vehicle industry, the disposal industry and the authorities, are represented, drafts a Technical Directive on "Motor-vehicle Recycling", which specifies and defines the requirements necessary for the approval and operation of a recycling establishment. Adherence to this Technical Directive is verified at regular intervals either by the responsible authority or by a private-industry institution.

- To permit even more comprehensive recycling, both the shredder process and other disposal routes, such as chemical recycling, metallurgical recycling, thermal recycling and also other new methods and processes currently still under development, will be approved.

Description of Concept

- **Reception points for life-expired motor-vehicles take the form of motor-vehicle trade establishments and certificated recycling establishments**
- **Last owner receives a recycling certificate.**
- **The residual value of the life-expired vehicle is negotiated between the last owner of the vehicle and the recycling establishment.**
- **Body representing all participants drafts a Technical Directive for Motor-vehicle Recycling.**
- **Shredder process and other disposal methods to be approved.**

Figure 1 Joint concept for motor-vehicle recycling

All parties involved must make their contribution and be apportioned rights and obligations in just measures in order to improve the motor-vehicle recycling situation described initially.

- The motor-vehicle industry will be responsible for incorporation and refinement of the latest developments in recycling-oriented design and for the development of a motor-vehicle recycling system suitable for vehicles currently on the road.

- The vehicle owner must undertake to use the vehicle in a recycling-oriented manner and to surrender the vehicle to one of the approved acceptance points when its useful life has expired.

- The recycling industry will concentrate on the draining, dismantling, reuse and recycling of components and materials.

- The basic and raw materials producing industries must be prepared to pursue jointly with the motor-vehicle industry the development of rational methods for the recycling of used materials from life-expired vehicles and to actually reuse these materials.

The support of the confederations and associations affected has helped in achieving the following agreements from the motor-vehicle industry:

- The setting-up of a take-back and recycling system, coupled with an assurance that each vehicle will be accepted back at a freely negotiable price.

- Increase in the materials recycling rates by 20 % of the residual materials currently not recyclable by the year 2000 and by a total of 60 % as a further objective, making allowance for recycling processes, the potential of which has not yet been fully realized.

- The setting-up of a take-back and recycling system within three years of the creation of the necessary background conditions by the legislature.

This modified "joint concept" was presented to the public and the relevant ministries in the spring of 1993 as an alternative proposal to the Federal Minister of the Environment's August 1992 Draft Ordinance on the Avoidance, Reduction and Recycling of Waste from Motor-vehicle Disposal.

Salient points

- o **Free-of-charge acceptance-back of life-expired vehicles and spares by the manufacturer or the marketer.**
- o **Setting-up of a universal network of recycling establishments.**
- o **The recycling establishments' suitability must be verified by means of inspectors' reports.**
- o **Introduction of a recycling certificate.**
- o **Obligation to recycling-oriented design of new vehicles.**
- o **Quotas for materials recycling of dismantled/recovered materials with obligation to annual verification to the responsible authority.**

Figure 2 BMU draft ordinance (August 17, 1992/June 3, 1993)

The salient points of the draft ordinance of August 17, 1992 and of the June 3, 1993 revised version can be summarized in the form of the following central statements (Figure 2):

- free-of-charge acceptance back of life-expired vehicles and spares from the final owner by the manufacturer or the recycling establishment operator;

- the setting-up of a universal network of take-back points and recycling establishments by the manufacturer or marketer; the number of take-back points must be the same as the number of marketing outlets;

- the manufacturer must submit an inspector's report regarding the suitability of the recycling establishment;

- introduction of a recycling certificate;
- obligation to recycling-oriented design of new vehicles;
- target materials recycling rates for the steel, nonferrous metals, glass, plastics and elastomer materials groups contained in vehicles, and for tyres; complete with requirement for annual verification to the responsible authority.

The German motor-vehicle industry rejected the present version of this draft at the BMU hearing on October 23, 1992, with the following argumentation:

- "Free-of-charge acceptance-back" is rejected, since it is an infringement of basic market-economy principles.
- A network of take-back points of the same density as the marketing network guarantees neither locational planning suitable to the problem nor the future economic viability of the recycling establishments.
- Documentation by accredited inspectors that disposal establishments fulfil the requirements of the Waste Act contradicts public-law principles. It would transfer state obligations to the motor-vehicle manufacturers.
- The recycling objectives stated for re-utilization and for materials recycling are not realistic.
- The recycling certificate recommended would give rise to a level of bureaucratic complexity completely improportionate to the effect desired.
- The creation of nation state, protectionist hindrances to trade in the EU.

Despite this extremely comprehensive rebuttal by the motor-vehicle industry, which was joined by the Confederation of Motor-vehicle Importers and of the Motor-vehicle Trade and the whole of German industry represented by the BDI (Confederation of German Industry) and the simultaneous development of market-economy alternatives in the form of the joint concept, the Federal Ministry of the Environment initiated inter-departmental consultations in November 1993, on the basis of the revised June 3 draft.

Status of life-expired vehicle recycling at Adam Opel AG

In pursuance of the original VDA concept and the joint concept, Adam Opel AG has formally accepted responsibility for the future life-expired vehicle recycling of its products and integrated this responsibility into its corporate strategy.

Principle points of emphasis are:

Recycling-oriented design
Use of recyclates,
The setting-up of a recycling network.

Recycling-oriented design

Whereas technical and economic factors, such as the load-bearing capacity of a component or material, and the associated manufacturing costs, have previously been the definitive element in design, they are now joined by assessment of subsequent recyclability and disposability of the individual parts and components.

Increasingly great significance will attach to the recyclability of an assembly or component under competitive aspects, above all in products manufactured on a mass-production basis.

Recycling-oriented design guarantees the recyclability of our products and promotes the economic viability of the recycling process. It is, in addition, the basic prerequisite for the achievement of a high recycling rate and therefore constitutes a significant contribution to the conservation of the environment, in terms of relieving the load on landfills and conserving raw materials and energy resources.

The question of the recycling and disposal of a product must be addressed as early as the development phase: Recycling starts on the drawing board.

Consistent achievement of recycling-oriented design in our new projects is therefore one of Adam Opel AG's central development objectives.

As we see it, the essential criteria are:

- design-stage provisions intended to ensure efficient removal of all operating fluids at the recycling stage,

- the selection of recyclable and environmentally safe materials,

- easy and therefore cost-effective dismantling of all components which are to be removed for the purpose of recycling from the future life-expired vehicle and

- the use of recyclates in suitable components.

At Opel, work on the development of suitable proposals and recommendations for the above topic was started at an early stage, in order to provide the designer with support in this area. The result is the "Recycling-oriented Design Manual" (see Figure 3), which deals in detail with the main criteria defined above.

An important precondition for environmentally safe disposal of life-expired vehicles is the most complete possible removal of operating fluids and fluid-bearing components, so-called "drainage".

The operating fluids used in motor-vehicles are, practically without exception, potential soil and groundwater pollutants.

Technical Development Center Europe
B&E Scrap Vehicle Management

DESIGN FOR RECYCLING

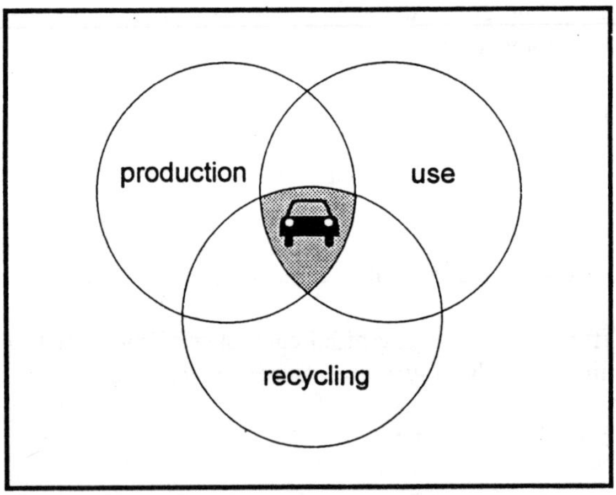

First edition

Figure 3 Recycling-oriented design manual

The manual provides the designer with suggestions for the incorporation into his new designs of suitable provisions to permit rapid removal of residual fuel, engine and gearbox oils, brake fluid and coolant, refrigerants and hydraulic fluid.

The heading "Recycling-oriented Materials Selection" is concerned with the development of a component which, at the point of the designer's decision on the material to be used, is optimum in terms of function, quality, cost-effectiveness, logistics and environment.

Obviously, to achieve recycling-oriented materials selection, the recyclability of the material must be assured. The target should be recyclability at a high value level.

In addition, the use of the smallest possible number of different plastics is desirable. The greater the number of different types of plastic, the more complex subsequent recycling will be, since the sorting necessary inevitably has a direct detrimental effect on the economy of the recycling process as a whole.

Assembled components consisting of different (i.e., chemically incompatible) polymer materials should be avoided. Materials compatible combinations should, therefore, be preferred.

The final two criteria for recycling-oriented materials selection are the obvious "Avoidance of Hazardous Materials" and the "Marking of Components" with information on their materials group.

The next topic is "easy dismantling" and "dismantling-oriented design".

The basic precondition here is, of course, ready accessibility of the components to be removed or dismantled and their rational dismantling with, wherever possible, no need for special tools.

In addition, vehicle design should devote attention to ensuring that appropriate configuration of structural components and assemblies facilitates subsequent dismantling operations.

Plastic components should, furthermore, be designed without reinforcing metal inserts and type-pure, i.e., single-material assemblies and assemblies given preference.

Where quality and functionality reasons dictate that no type-pure design is possible and the materials used are also not mutually compatible, the design must provide ready separability by selecting an appropriate joining method. The bumper design for the new Opel Corsa provides a good example of dismantling-oriented design (see Figure 4).

The bumper consists of the bumper mounting and the bumper surround. Division of the two functional components has been achieved in this design, i.e., there is no fixing-element-based joint between the bumper and the mounting.

The bumper surround is produced from modified polypropylene and is clipped to the bodyshell and fixed using only two screws to the two side wheel arches. The bumper mounting consists of steel and is a bodyshell component and is bolted to the bodyshell.

For the purpose of dismantling, the bumper can be quickly and simply torn out of its clips or off of its screws. The steel mounting remain in the bodyshell.

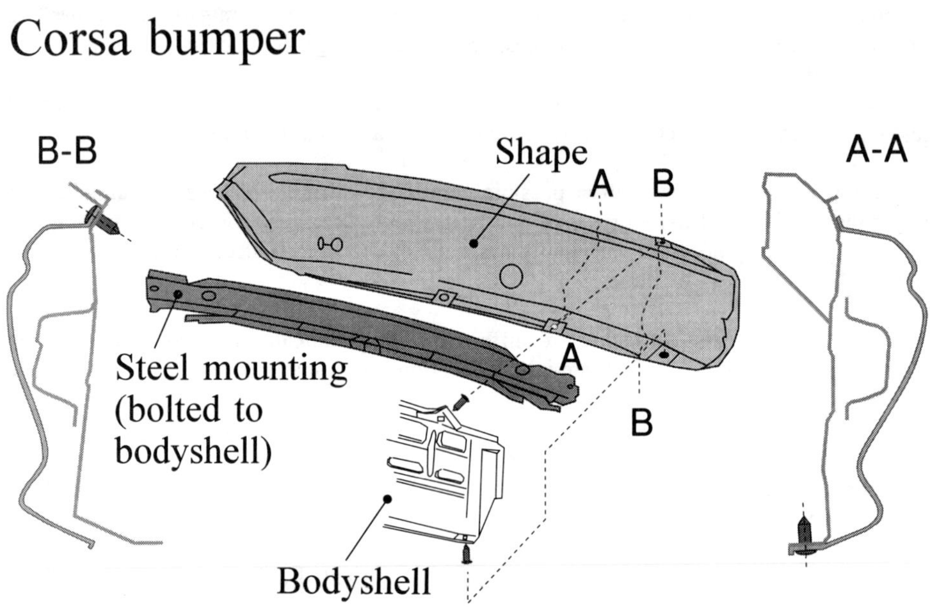

Figure 4 Dismanteling-oriented design

The reuse of recycling material is also a target and a further criterion of recycling-oriented design.

At Adam Opel AG, we have concentrated on the use of recyclates and on the creation of complete materials cycles.

Materials cycles

Polypropylene		
Battery casings Bumpers	⇨	Wheel-arch lining Bumper mountings Air-filter housings
Polyurethane		
Seat foam	⇨	Insulating mats
Polycarbonate		
Bumper	⇨	Spoiler mountings
SMA		
Insert in instrument panel	⇨	Air-duct in instrument panel

Figure 5 Use of recyclate

In the definition of the plastics materials cycles achieved up to now, it has been attempted to develop and use not only components with various requirement profiles, but also various plastics types and processes up to readiness for series production (see Figure 5).

The prime priority was to create a complete cycle for polypropylene. The recyclate obtained from life-expired battery casings and bumpers is now reutilized in wheel-arch linings, bumper mountings and air-filter housings.

Polypropylene was targeted for a number of reasons. It is, on the one hand, the most heavily used material in the new designs (the Vectra, Calibra, Astra and Corsa, each contains around 40 kg of this material) and, on the other hand, it can be easily modified to achieve required specifications. In addition, polypropylene can also be reprocessed several times without significant losses of properties. The second cycle was created for polyurethane. Flexible polyurethane foam is available as upholstery padding in large quantities both at vehicle scrapping and during manufacturing (production waste). The development objective in this case was the creation from this waste material of insulating mats which could be used for noise suppression purposes.

For this purpose, the flexible polyurethane foam is firstly comminuted to foam flocks and then processed to achieve a noise-insulating mat using a new polyurethane-based binder. Road tests have confirmed that the required attenuation figures are achieved when this material is used for firewall insulation. This mat has been in series use for two years now.

There is another complete material cycle for one other materials group, the polycarbonates. This makes it possible to recycle the painted bumpers of the Opel Senator and Kadett GSi.

The material, a polycarbonate/polybutylene-terephthalate blend, is used after reprocessing to produce roof-spoiler mountings for the Opel Astra and the new Corsa GSi.

And, finally, as early as late 1992, we had created a closed materials cycle for the plastic SMA, a modified styrene. In this case, large quantities of die-cutting and punching waste occur as production residues (up to 40 t die-cutting and punching waste per month) during production of the instrument panel for the Opel Vectra.

The combinations consisting of three completely different types of plastics, namely SMA insert, PUR foam and ABS foil, are separated into three fractions at our development department's establishment using a special physical/mechanical technique known as the "hydrocyclone process". The SMA fraction, a glass-fibre-reinforced modified styrene, is recycled and used in the production of air ducts for the Vectra instrument panel. It goes without saying that these recyclate applications fulfil all the specifications applicable to a new product.

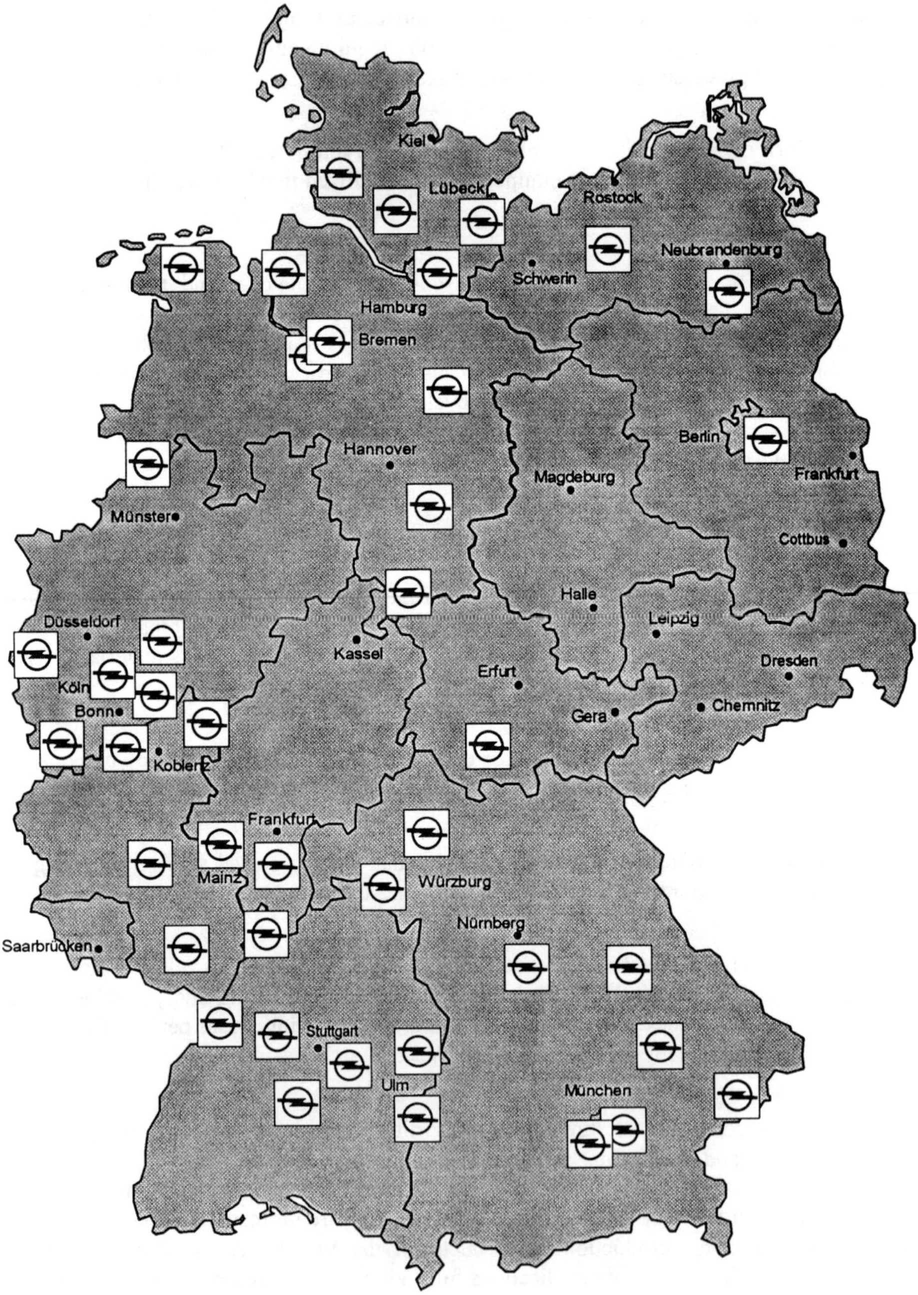

Figure 6 Network of recycling establishments

Setting-up of a recycling establishment network

The continuous provision of used plastics is, of course, the indispensable precondition for the maintenance of the above-mentioned materials cycles. In addition, Adam Opel AG regards the setting-up of an efficient network of life-expired vehicle dismantlers and the reinforcement of the existing network as a company responsibility and a contribution to the assurance of the future environmentally safe disposal of all life-expired Opel vehicles and vehicles of other makes.

In order to achieve this, it was firstly necessary to draw up a generally recognized requirement profile for the equipping and procedural structuring of life-expired recycling establishments.

This was drafted and coordinated within PRAVDA

The criteria contained in it constitute, on the one hand, Adam Opel AG's assessment criterion for the recycling establishments to be certificated and, on the other hand, they are intended for incorporation as a foundation for a standard federal provision in the Technical Directive for Motor-vehicle Recycling proposed in the "Joint Concept".

The fulfillment of the criteria bindingly prescribed in the requirement profile will ensure that only dismantlers who work in an environmentally safe and responsible manner are able to obtain a cooperation agreement with Adam Opel AG and will subsequently also be empowered to issue a recycling certificate.

Work continues on the expansion of the recycling establishment network in Germany. In late 1993, Adam Opel AG was able to certificate forty-two dismantling establishments (see Figure 6), i.e., these establishments have undertaken to work under market conditions in accordance with the agreed environmental criteria. This is a service for the final owner. These recycling establishments are, of course, also permitted to cooperate with other motor-vehicle manufacturing companies, in order to ensure the economic basis for their activities. We are endeavouring to draw this net even tighter, in order to prevent final owners having to travel long distances and in order to ensure a fair and competitive price for their life-expired vehicles.

Adam Opel AG's commitment to the implementation of the market-economy oriented "Joint Concept" is real and is demonstrated by the companies attention to recycling-oriented product design, the creation of closed materials cycles and the setting-up of a take-back and recycling network. The fact that these activities have been accomplished despite the lack of the legal background conditions long necessary, such as the recycling certificate, for instance, proves that the motor-vehicle industry is ready and able to assume responsibility for the return and recycling of life-expired vehicles and is performing the associated tasks with a view to achievement of the objectives.

Automated material identification in recycling

Derek Gentle

Introduction

The automotive industry faces worldwide pressure to help find environmentally friendly ways of disposing of end of life vehicles. Most national governments have some form of proposed legislation to cover vehicle disposal and the EU have designated vehicle scrap as a priority waste stream. It is not yet clear if legislation will impose targets or what those targets will be. There are questions of economic viability, environmental balance, and lowest energy use which must be considered, alongside the necessity to satisfy other legal and customer requirements.

Cost estimates for recovery and recycling of plastics material tend to ignore the time and cost associated with identifying the material. Most dismantling studies start with a knowledge of the material in the parts to be removed. In practice the cost of material identification by non-automated systems, such as referring to dismantling manuals or finding and reading material identification marks, may well tip material recycling into the non-economical category.

Mistakes in material identification can also have serious consequences on the later stages of the recycling process, where contamination by non compatible polymers can ruin whole batches of recyclate.

The development of automated material identification equipment was considered of high priority. This paper reports on the successful development of two different types of identification equipment and discusses some of the other techniques evaluated during the development process.

Background

Although cars are already recycled to a greater extent than any other consumer product, most manufacturers have agreed to improve the recycling capability of their vehicles. The metals in end-of-life cars are already recycled, therefore the non-metallic content provides the only opportunity for improvement.

Ford's interest in automated identification processes started in 1989 when a project, to identify plastics materials in shredder residue on the conveyor belts leaving the shredder, was funded at Southampton University. It was considered that if plastics could be treated as another material stream recovered from the shredder the recycling would be economically viable.

Most modern cars contain between 21 and 23% of solid non-metallic materials, this makes up the Shredder Light Fraction which presently goes to landfill during vehicle disposal.

Plastics make up about half of this, or about 12% of total vehicle weight, but this includes fabrics, resin bonded products, etc. Only about 7% is moulded plastics.

The use of plastics has allowed vehicle safety, security, comfort, performance and feature improvements to be achieved without excessive weight increases. The energy and resource consumption for building a car are on average less than 10% of those used to run the vehicle. Recycling for the sake of recycling does not ensure that resources are conserved or that pollution is minimized. If materials can be recycled economically, without using more total energy than they contain, then they should be recycled, but if not then the energy should be recovered.

Because we presently do not know what can be achieved in reducing the total recovery and processing costs for plastics, or where the cut off point for economic recycling will be, we can not say that energy recovery is the only answer. Material recycling can not be ignored. As long as we are not certain what is economically viable and what may be required by future legislation we have to design cars and use materials which give the best opportunities for recycling.

At the present state of technology it must be assumed that any plastics recovered must be from component dismantling rather than from shredder residues. Ford is one of the vehicle manufacturers that have established pilot in-house dismantling facilities, as well as partnerships with dismantlers, aimed at understanding the costs involved in such a recovery process. Close cooperation by the automotive industry with raw material suppliers and recyclers have provided substantial, but still incomplete, information on the economics of plastics recycling via dismantling and the potential for recyclate use in new cars.

A major shortcoming of the cost statistics generated from dismantling trials is that no allowance is made for the time required to identify the materials recovered. Components dismantled in such controlled trials are predefined and the materials known. If the time taken to identify materials by existing methods was added to the recovery cost even less material recycling would be economically viable.

It became obvious that some of the identification techniques which were being developed for plastics in shredder residues could be modified to identify materials at the dismantling stage. This was seen to be a much shorter term and easier target than shredder residues, therefore Southampton were directed to concentrate on identification systems for use by dismantlers and material recyclers.

Two different techniques have been developed by Southampton on behalf of Ford. One, based on medium frequency infrared spectroscopy has some common features with the equipment under trial at BMW Landshut on behalf of the JRC, the other is a completely novel technique based on electrostatics which is completely portable and free from potential operator safety issues.

Dismantling

Dismantling studies have indicated that approximately 20 man minutes of dismantling time can yield about 30 kg of thermoplastics components at a dismantling cost of about than 1 DM/kg. A further 10 min. can yield about 20 kg of PU foam and RIM. Although this varies from one vehicle to another, similar results are reported by different manufacturers. Figure 1 illustrates how the yield drops off with time and Figure 2 shows how rapidly cost of removal increases with time as the available components become smaller and more difficult to remove.

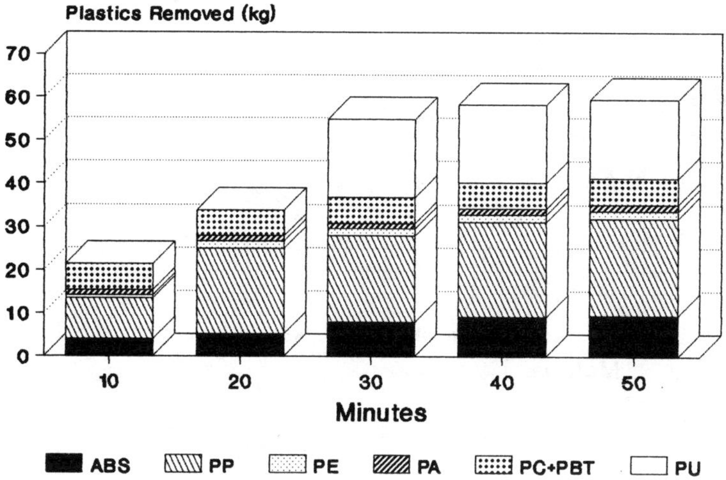

Figure 1 Yield of dismantled plastics/time

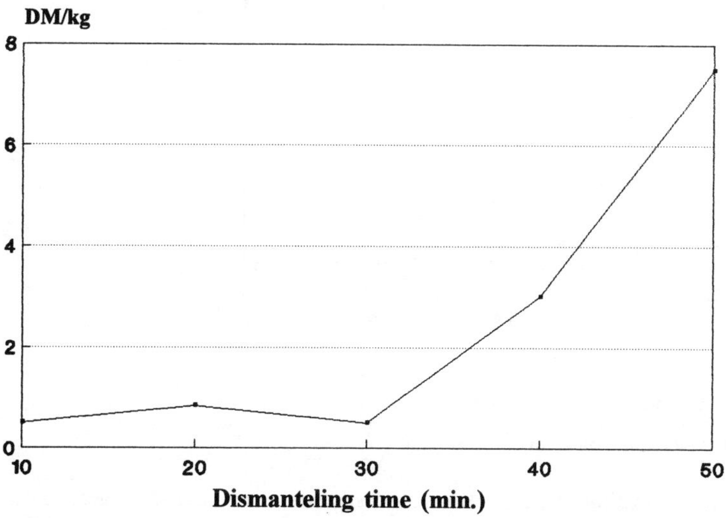

Figure 2 Dismantling cost of plastics/time

Taking 1 DM/kg as the dismantling cost, the total cost of recyclate is projected as being between DM2.5 and 3.5/kg, depending on the necessity to separate, delaminate, or remove paint, etc. Note that this cost is virtually independent of the original raw material price.

Figure 3 shows the comparison of this potential recyclate cost against the costs of major plastics types. This illustrates the how difficult it is to justify material recycling on economic grounds.

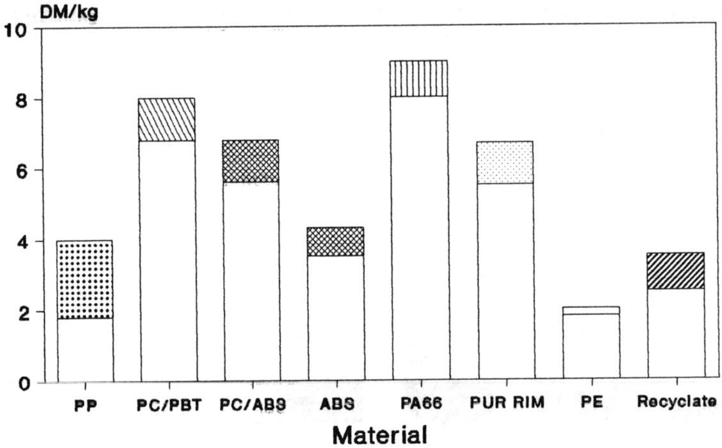

Figure 3 Recyclate v. virgin plastics cost

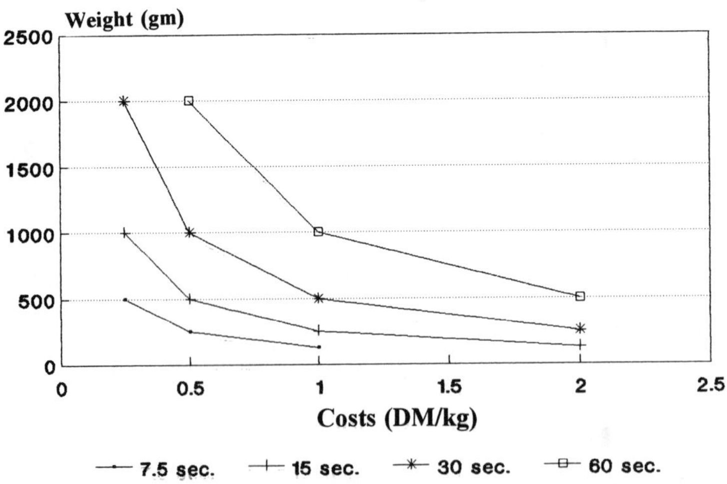

Figure 4 Effect of identification times on recovery costs

Conventional material identification systems rely on the dismantler using manuals, fact sheets, computer programs, or looking for the materials identification code on the part. An indication of the costs involved in this process is obtained from Figure 4, which also clearly indicates how dismantling of small components is most impacted by the additional cost.

By developing better, faster methods of identifying materials the dismantling process can be designed along production line principles, with plastics ripped from the body with no concern as to sorting. Separation into material types can then take place on a conveyor belt system, perhaps at a central recycling facility. This would also tend to reduce the risk of cross contamination with other polymers, and separate recycling streams for minor constituents such as ABS/PC or POM could become viable.

Potential spectroscopy identification processes evaluated

Objectives of the Ford development program were to develop equipment for identifying the different plastics materials used in vehicle production under the poor conditions found at most dismantling sites. The equipment should be accurate, robust, fast enough to keep the variable cost of its use as low as possible, easy to use by unskilled workers, and preferably portable.

Because many of the plastics used on cars are black in colour, and may be heavily filled, some methods such as Near Infra Red Spectroscopy which can not identify black samples were ignored.

TIRES
Much of our initial efforts were using TIRES (Transient Infra-Red Emission Spectroscopy) which had been reported in the literature as capable of identifying plastics materials even when pigmented black. The theory is that by only heating the surface of a bulky sample an emission spectrum of the sample occurs before black body emission takes place. The laser beam used to obtain the rapid heating must move constantly over the surface so that the sample does not burn. Even by spinning the sample we could not achieve the control necessary to consistently identify an emission before black body radiation or burning took place.

Having spent several months trying out different lasers and other heating methods we concluded that the TIRES method, even if it could be made to work consistently would be too slow, too expensive, environmentally sensitive, and possibly be dangerous in uncontrolled conditions.

Raman spectroscopy
Worked with light coloured samples but dark colours gave poor spectra and black samples tended to burn.

Laser decomposition
The decomposition products of laser heating were analyzed by gas phase infrared spectroscopy. Apart from PVC the spectra tended to be the same for all polymers along with

black body emission associated with bulk heating. (JRC has apparently achieved better result with an experimental unit). We did not consider this technique worth pursuing due to concerns raised about high power laser usage in a dismantling environment.

DRIFT spectroscopy

Diffuse Reflectance Infrared Fourier Transform spectroscopy is a technique commonly used for the analysis of opaque powder samples. An infrared source radiates the sample and the radiation reflected from it is analyzed. Good spectra were obtained, particularly from lightly abraded samples. One interesting technique which was evaluated in the early stage of the development is to use silicon or aluminium oxide abrasive papers to abrade the sample surface, the powder trapped on the abrasive paper can the be analyzed by inserting the paper in the DRIFT cell. Rover have recently installed a standard DRIFT equipment and we understand are using the abrasive paper technique and standard spectra to identify unknown materials in their dismantling trials.

A related method is **SR (Specular Reflectance) spectroscopy** which proved to be equally successful at identifying all the different polymers and blends. The infrared spectra are essentially the same for both DRIFT and SR but care must be taken to separate the diffuse and specular reflectance as the spectra are different shapes and can interfere. Again the silicon carbide paper sampling technique can be used with SR spectroscopy. Figure 5 illustrates the difference between DRIFT and SR spectroscopy.

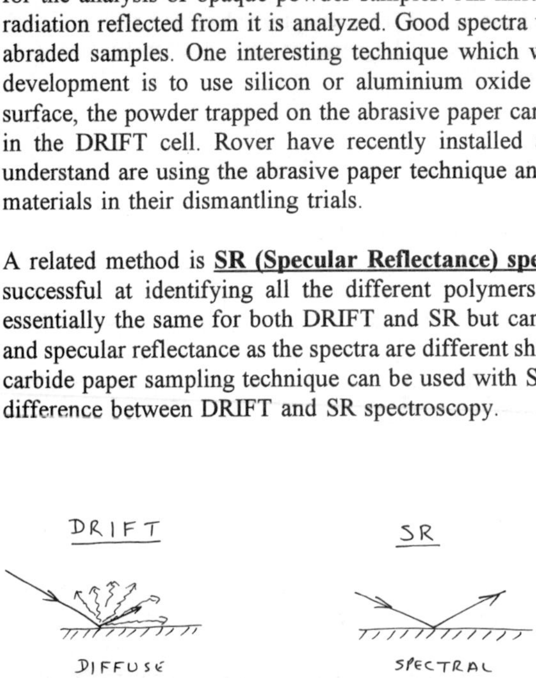

Figure 5 Principle of DRIFT and SR spectroscopy

Both of these systems use **Medium wavelength InfraRed (MIR)** and worked on all colours, with fillers, and with grained surfaces. Some fillers such as talcum do show up on the spectrum and this effect must be used and separated from the polymer spectra the computer system. Identification times about 5 seconds were possible.

The effectiveness of the system can be seen from the three spectra shown in Figures 6, 7, and 8 which show PC (polycarbonate), PBT (Polybutylene terephthalate), and the PC+PBT blend Xenoy used in Ford bumpers.

It is possible to identify the products of weathering on the samples, this does not interfere with identifying the base polymer but could be used to eliminate badly degraded materials from the recycling stream.

Having decided that MIR gave the best chance of developing a satisfactory spectroscopic instrument for use in dismantlers, or more likely in centralised separation facilities at recyclers because of the probable cost of the equipment. Our objective was to develop a robust equipment which could be located in a dismantlers or car breakers facility.

Ford have developed with Southampton University a material identification system which allows rapid identification of materials in a recycling environment. SR spectroscopy was chosen as the basis for the development. It was found that picking out specular reflectance without diffuse reflection interference in the SR mode was easier that filtering out specular reflectance from diffuse reflection in the DRIFT system, especially when samples had a grained surface.

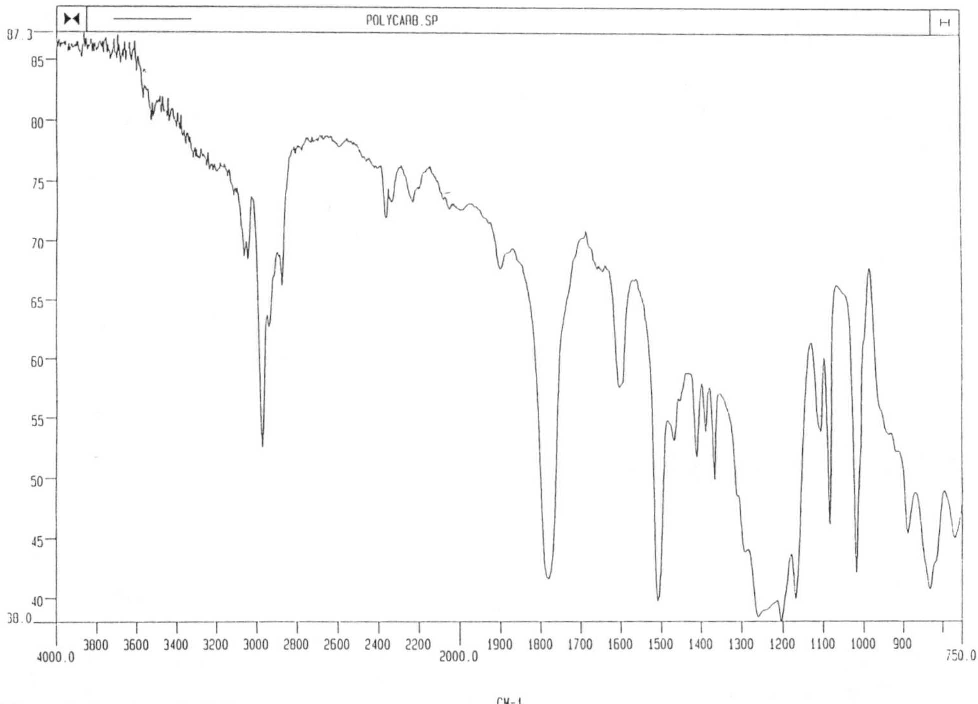

Figure 6 Spectrum for PC

Identification equipment developments

Taking samples to put in a cell, even using the silicon carbide paper technique, was considered to be unrealistic because of time and problems with contamination of the equipment. The equipment should be totally enclosed.

The most sensitive equipment uses liquid nitrogen cooled sensors, although this may be practical in large well run recycling facilities, it was felt that liquid nitrogen sensors should not be used.

Southampton were asked to develop an optical system which would enable the equipment to be completely enclosed and the sample held to a so called "window to the world", that is a aperture closed with a scratch resistant window. This optical system to be attached to an enclosed self contained MIR spectroscopy system connected to a computer loaded with all the spectra likely to be found from plastics used in cars. Obviously there is a balance between operating speed and the number of spectra to be interrogated by the computer, the target set was to identify all common thermoplastics and blends, and SMC, PUR, and phenolic. More specific computer programs to separate different PUR types etc. would be developed at a later stage.

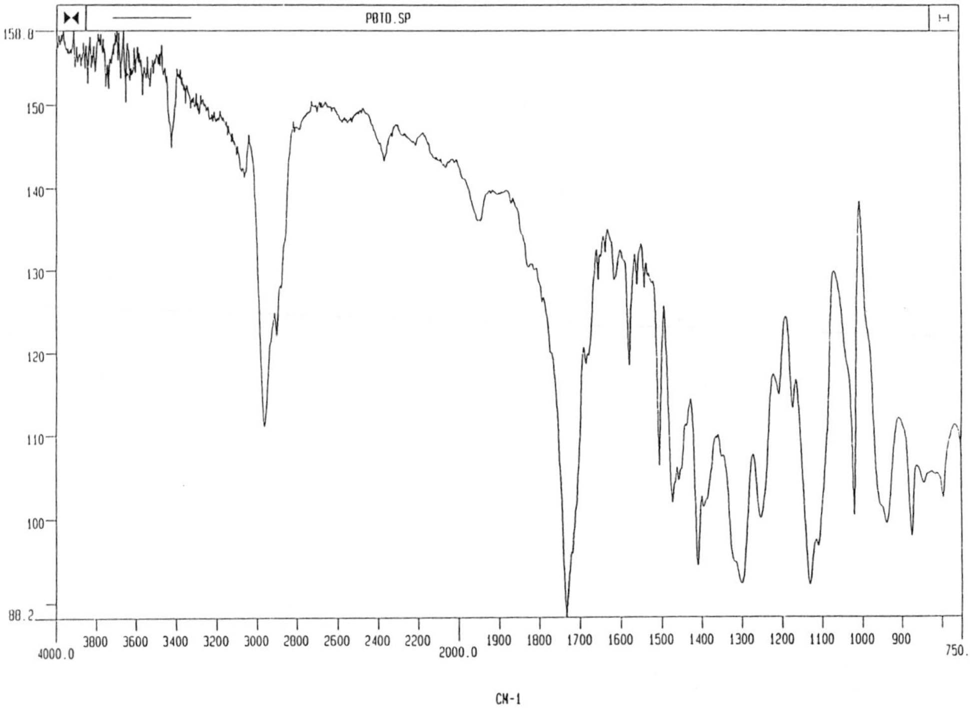

Figure 7 Spectrum for PBT

Southampton have successfully built prototype systems which have fulfilled all our objectives. Although the use of external optics and uncooled sensors reduce the sensitivity, the computer analysis of the spectra allows in the region of 90% positive identification and no mistaken identification. Further development of the computer analytical method may even improve this result by eliminating the interference from talcum.

Ford's intention is to demonstrate the feasibility not to build equipment. We are working with manufacturers of Infrared spectroscopy equipment on the development of production equipment. It is planned to install the first production unit in our Pilot Dismantling Facility in Cologne. It is very satisfying that the JRC program came to the same conclusion as to the spectroscopy technique which should be followed. Combining resources to get the best of all systems is an obvious next step. Making equipment available for improving and lowering the cost of recycling plastics may be one of the most constructive contributions that the car manufacturers can make.

The shortcomings of MIR spectroscopy is that the sample needs to be taken to the equipment, unless silicon carbide paper is used, and the equipment is rather expensive for a small dismantler to install. It is ideal for segregation of materials at central recycling facilities, and could save duplication of identification effort at dismantlers and recyclers.

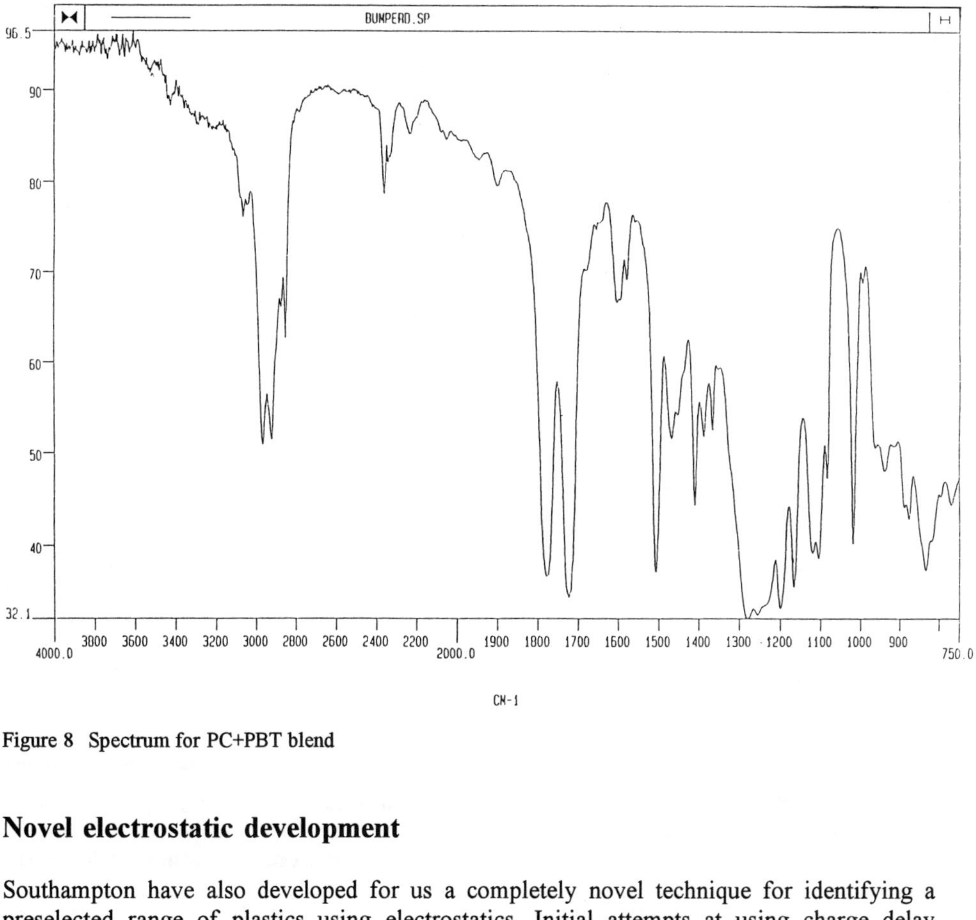

Figure 8 Spectrum for PC+PBT blend

Novel electrostatic development

Southampton have also developed for us a completely novel technique for identifying a preselected range of plastics using electrostatics. Initial attempts at using charge delay appeared to be promising. A high voltage corona source was used to electrostatically charge the samples. The rate at which the charge on the grounded sample dissipated was measured on an electric field meter coupled to a chart recorder. Although these trials did show differences in decay rates for different plastics they were affected by the type of surface finish and were therefore unreliable for our type of components.

This early work did encourage the continued study of electrostatic principles. Going back to historic work on electrostatics the Triboelectric Series was examined. This old principle of unlike materials developing opposite charges when rubbed together, such as cats fur and ebonite rods, appeared to warrant further investigation. The series is an empirically derived list of materials, including metals, plastics, natural products, ceramics, etc. Materials which develop positive charges are always above those which cause the charges when the two are rubbed or scraped together. The series was very incomplete as previous workers had apparently concentrated on limited areas such as ceramics or metals. Our first priority was to expand the series to include the major plastics.

The technique which has been developed uses 'charging head' materials selected from strategic known points on the Triboelectric Series and bringing them into frictional contact with the plastic under test. Variations in the polarity and/or magnitude of the resultant electrostatic charge generated on the heads would indicate the position of the unknown plastic in the Triboelectric Series.

After considerable development effort a completely portable, battery powered, prototype equipment has been completed. The present prototype works entirely on polarity differences and can separate four preselected families of plastics. By using charge magnitude it is possible to increase this to identify eight different materials. Selecting the head materials allows the selection of materials to be varied. Figure 9 shows a typical selection of head materials and the typical plastics which can be separated.

PLASTIC SAMPLE	HEAD MATERIAL		
	Acrylic	Brass	PVC
PA, POM, PC, PUR	-	-	-
ABS	+	-	-
PP	+	+	-
PE	+	+	+

Figure 9 Head selection for Triple Head Triboelectric Identification

It was found that metals fall close together in the series so only one head can be metal, brass as been used in most of our experimental work.

The hand held prototype meter has given even better results than expected, giving very accurate results during trials on a variety of old car parts.

This low cost portable equipment is considered to be ideal for pre-selection of components at the dismantling stage. For example identifying between the limited ranges of materials used in bumpers, instrument panels, wheel covers, air cleaners, and interior trim, is fairly easy. An integral paint stripping head would be fairly simple to incorporate into the equipment.

Discussions with potential manufacturers are underway and the first production units will hopefully be available by the end of the year. Meanwhile prototype units will be used in trials at selected dismantlers.

Apart from being portable this equipment has the advantage of almost instantaneous results, no umbilical cords to stationary equipment, very light weight, manoeuvrability, and no dangerous elements such as lasers or high voltages. Head cleaning when necessary is very easy.

Conclusions

Although energy or feedstock recovery is a highly desirable alternative to material recycling there are still pressures on the vehicle manufacturing industry to use materials which can be recycled.

The MIR spectroscopy and Triboelectric techniques and equipment developed by Southampton University for Ford can contribute to the automotive industries effort to make material recycling more economically viable. Various patents have been applied for as part of this development program. These are aimed at protecting Ford and the rest of the industry from any attempt, by others, to limit our access to identification technology for commercial ends. We look forward to working with JRC and other industry groups to encourage further evaluation and implementation of the technology by the disposer and recycling industries.

Acknowledgments

The author would like to thank the Ford Motor Company for permission to use results from their recycling work in this paper, and JRC and Rover for permission to refer to their work.

List of Authors

Axel Balk , Volkswagen AG, Abt. ENK
38436 Wolfsburg, Germany

Rainer Bickel , Gottlob Auwärter GmbH + Co
Postfach 81 01 40, 70518 Stuttgart, Germany

Prof. Dr.-Ing. Peter Eyerer , Universität Stuttgart, Institut für Kunststoffprüfung und Kunststoffkunde
Postfach 80 11 40, 70511 Stuttgart, Germany

Derek F. Gentle, Ford Motor Company Ltd., Room 15/363
Waindon
GB-Basildon, Essex SS15 6EE,

Hans-Günther Haldenwanger VDI, AUDI AG, Abt. I/EXW
Postfach 10 02 20, 85045 Ingolstadt, Germany

Dr.-Ing. Jens Hesselbach, PE Product Engineering GmbH
Friedhofstraße 1, 73760 Ostfildern, Germany

Guido Horsch, Dr.Ing.h.c.F. Porsche AG, Abt. EGL
Postfach 11 40, 71283 Weissach, Germany

Lothar R. Jung, Ymos AG Industrieprodukte, Abt. Lackentwicklung
Postfach 22 40, 63171 Obershausen, Germany

Bernd Kiefer, Adam Opel AG, Abt. B & E Fahrzeug-Recycling
65423 Rüsselsheim, Germany

Dr. Winfried Köppen, Volkswagen AG, Abt. E/FS-RA
38436 Wolfsburg, Germany

Bob Lee, Gottlob Auwärter GmbH + Co,
Postfach 81 01 40, 70518 Stuttgart, Germany

Dr.-Ing. Oswald Parr, Filterwerk Mann & Hummel GmbH
Hindenburgstraße 37 - 45, 71638 Ludwigsburg, Germany

Reinhard Pälmer, Wilhelm Karmann GmbH, Abt. Kunststofftechnologie
Postfach 26 09, 49084 Osnabrück, Germany

Ingrid Pfleiderer, Universität Stuttgart, Institut für Kunststoffprüfung und Kunststoffkunde
Postfach 80 11 40, 70511 Stuttgart, Germany

Herwig Reim, AUDI AG, Abt. I/EXW
Postfach 10 02 20, 85045 Ingolstadt, Germany

Herbert Rösch VDI, Karl Kässbohrer Fahrzeugwerke GmbH
Kässbohrerstraße, 89077 Ulm, Germany

Konrad Saur, Universität Stuttgart, Institut für Kunststoffprüfung und Kunststoffkunde
Postfach 80 11 40, 70511 Stuttgart, Germany

Dr.-Ing Siegfried Schäper VDI, AUDI AG, Abt. I/EKW
Postfach 10 02 20, 85045 Ingolstadt, Germany

Ulrich Schatton, Mercedes Benz AG, Werk Sindelfingen, Abt. Lacktechnik
71059 Sindelfingen, Germany

Manfred Schuckert, Universität Stuttgart, Institut für Kunststoffprüfung
und Kunststoffkunde
Postfach 80 11 40, 70511 Stuttgart, Germany

Peter Schüller, Karl Wörwag GmbH & Co KG, Abt. Lackentwicklung/ Anwendungstechnik
Postfach 40 09 69, 70409 Stuttgart, Germany

Peter Straßer , Bayerische Motoren Werke AG, Sparte Kunststoff,
Abt. Verfahrensentwicklung
Ohmstraße 2, 84030 Landshut, Germany

Dr. Wolfgang R. Wittig, Mercedes-Benz AG, Zentralwerkstofftechnik,
Abt. Polymere (ZWT/KE), H 120
70322 Stuttgart, Germany

Dr.-Ing. Bernd Woite VDI, Bayerische Motoren Werke AG, Sparte Kunststoff/Textil
Abt. SK-5
Ohmstraße 2, 84030 Landshut, Germany

Dr.-Ing. Lizhou Yin, AUDI AG, Abt. I/EXW
Postfach 10 02 20, 85045 Ingolstadt, Germany

Peter Zumbroich, Adam Opel AG, Abt. B & E Fahrzeug-Recycling
65423 Rüsselsheim, Germany

Index